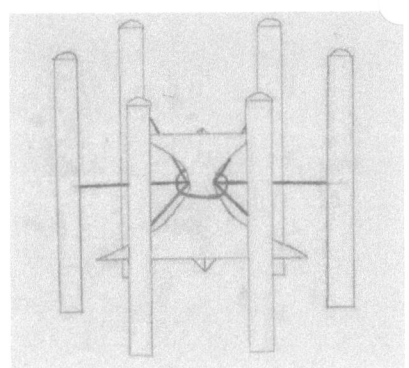

Multigenerational Starship Design Considerations: A Problem Based Learning Laboratory Experience

Edited by Harold A. Geller
with illustrations by Emma Rojas

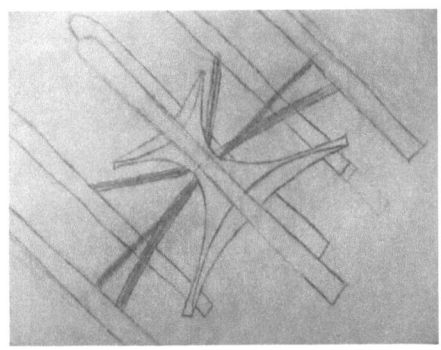

Multigenerational Starship Design Considerations: A Problem Based Learning Laboratory Experience

Edited by Harold A. Geller
with illustrations by Emma Rojas

Graphic Credits:
Front Cover – Emma Rojas
Back Cover – Harold Geller
Chapter Front Pieces – Emma Rojas
Editor Portrait – Emma Rojas
Illustrator Portrait – Emma Rojas
Others – Harold Geller or as noted

Table of Contents

Chapter 3: Starship Astronautics

Richard J. Oh

Chapter 4: Starship Political Infrastructure

Alisa Aydin, Peter Lam and Amy Li

Chapter 5: Starship Protection

Heather Gonyeau, Rachel Rockrohr and Katelyn Squicciarini

Chapter 6: Starship Resource Management

Katie Askegaard, Tim Betts and Kara Elser

DEDICATION

I dedicate this book to the memory of Dr. Donna Roudabush Sterling (28 August 1948 – 24 June 2014). Donna was raised in Sacramento, California. She received her doctorate in education from George Washington University in 1992. She joined the faculty at George Mason University in 1993. In 2013 she was named a George Mason University Distinguished Service Professor. She served as director of the Center for Restructuring Education in Science and Technology (CREST) and she was the Principal Investigator for the Virginia Initiative for Science Teaching and Achievement (VISTA). Donna received a letter of appreciation from President Barack Obama and on 12 June 2014 the Virginia General Assembly passed a resolution expressing appreciation for her commitment to the education of students in the Commonwealth of Virginia. I had the pleasure of working with Donna for the better part of a decade. It was Donna and her cohort, Dr. Wendy Frazier who introduced me to the Problem Based Learning (PBL) pedagogy. My utilization of the PBL in my Honors course in astrobiology served as the foundation for the development of this edited volume. I had the pleasure of serving as a co-Investigator and consultant on a number of science education grants led by Donna and Wendy, for whom I am grateful.

ACKNOWLEDGEMENT

I gratefully acknowledge the students of HNRT 228 Astrobiology, including, but actually not limited to: Juliane Marie Veloso, Allen James, Alisa Aydin, Peter Lam, Amy Yi, Heather Gonyeau, Katelyn Squicciarini, Rachel Rockrohr, Tim Betts, Kara Elser, Katie Askegaard, Amy Wynant, Tyler Durkee, Kristen DiMarino, and Ryan Vitter. There were a handful of students who declined to be a part of this publication effort. I also acknowledge Prabal Saxena, the graduate teaching assistant for my astrobiology course. I would also like to thank Dr. George Taylor, Dr. Kathleen Alligood, and Dr. Zofia Burr for allowing me to teach astrobiology as a special topics science course within the Honors College of George Mason University.

I would like to thank the late Dr. Donna Sterling (see dedication), Dr. Wendy Frazier, and Dr. Mollianne Logerwell for introducing me to the problem based learning (PBL) pedagogy, and allowing me to see PBL in action over so many years.

I also acknowledge Dr. John C. Evans, Emeritus Professor of Physics and Astronomy at George Mason University for his contribution to this volume; and, Dr. Joe Weingartner who reviewed a portion of this manuscript.

Finally, I would like to thank Thomas Jefferson High School participants in this effort including Sowmya Ranga, Sonia Thakur, and Richard Oh.

I am solely responsible for all modifications, additions, and deletions to the original manuscripts that were provided to me by the authors and reviewers.

PREFACE

This volume grew out of an effort within the Honors College of George Mason University during the spring semester of 2014. The special section in the Honors College from which these papers (excluding high school student participants) are derived was conducted by Dr. Harold Geller and his graduate teaching assistant Prabal Saxena.

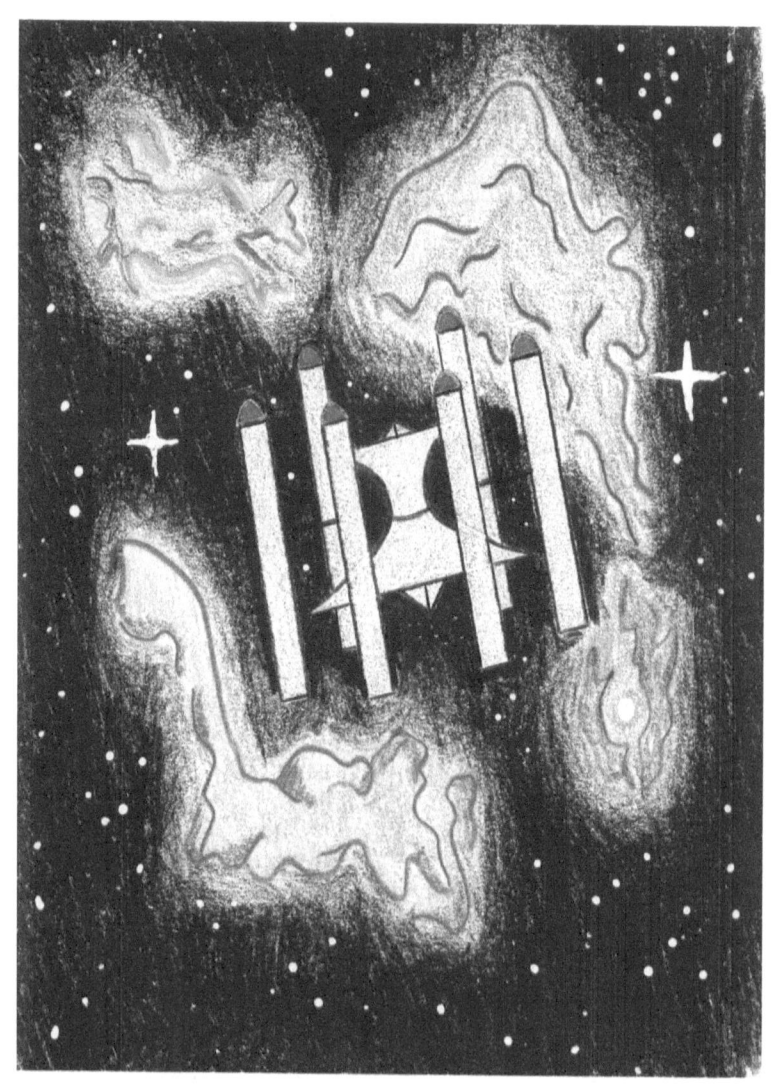

Introduction: A Course in Astrobiology
Harold Geller and Prabal Saxena

> A teacher should give his pupil opportunity for
> independent practice without suggestions from
> himself, and thus set upon him the stamp of
> indelible memory in its purest form.
> **Philo**
> ...and there is nothing new under the Sun.
> **Ecclesiastes 1:9**

0.1 INTRODUCTION

Astrobiology has been taught as a special topics course within the Honors College at George Mason University since the spring of 2002. In the fall semester of 2014, the manner in which the astrobiology course was taught was modified to include the problem based learning (PBL) pedagogy. (DeGraaff, and Kolmos, 2003; Frazier and Sterling,2008; Hmelo-Silver, 2004; Krajcik et al., 1998; Matkins et al., 2012; and Sterling and Frazier, 2010) We do not want our readers to believe that PBL is a "new" technique, as the quote from Philo above notes. Perhaps it is best described as a recent implementation of our wisest teachers over the millennium. We describe here our implementation of the PBL pedagogy that led to this volume.

The fall semester of astrobiology course consisted of 21 students. The course is a four credit course with a laboratory session that meets once a week for a 180 minute period, and lecture sessions that meet twice a week for a 75 minute lecture period.

The lecture period was run in what is today called the flipped pedagogy. The term "flipped" is not one of our own choosing. Both the faculty member of record and the graduate teaching assistant for this course are originally from the New York City metropolitan area, and are familiar

1

with a different connotation of the word "flipped."

For the lecture period, portions of the textbook by Geoffrey Bennett and Seth Shostak titled Life in the Universe, were assigned. The reading material was reviewed by the instructor, interspersed with a series of questions posed to the students utilizing a personal response system. The posed questions served as a formative assessment tool, and allowed the instructor to incorporate class participation into the lecture portion of the grading for the course.

After reviewing the material with the interspersed questions, students participated in group work, addressing a series of thoughtful questions. The groups consisted of 3 to 4 students wherein one student was charged to be the leader of the group, another student was charged to be the recorder of the response, and the other 1 or 2 students were active participants in coming to a consensus response to the question. The roles of the students were alternated from one meeting session to the next. At the end of the period, each group turned in one response to the series of questions, as recorded by the student who had been assigned the role of recorder.

In the fall of 2014 it was decided that the laboratory exercises, most recently accomplished by utilizing the Activities Manual for Life in the Universe by Prather, Offerdahl and Timothy Slater, should be implemented more in alignment with the problem based learning (PBL) pedagogy as implemented by Sterling and Frazier.

The Activities Manual (Prather, et al.) was still required for the laboratory sessions; however, in the fall of 2014 these exercises were "flipped" in the sense that the students performed the exercises in the manual on their own, and turned in their responses during the regularly scheduled laboratory session.

The laboratory sessions themselves, were the scene of the PBL implementation. The first session introduced the students to the problem at hand. In this case, the problem to be addressed in the lab during the entire semester was the design of a multigenerational starship mission to the stars. The students themselves were then given time to consider what questions they wanted to ask in order to come up with a multigenerational starship design.

It was always the intent of the instructors that the final product was to be multidisciplinary and not only address the science and engineering questions but also the sociological and psychological questions.

All students were asked to come up with questions they would need to address in order to design such a mission to the stars, and each question was to be written on a single three by five sticky note paper. The graduate teaching assistant (GTA) then stuck these notes on the whiteboards, organizing them as he saw that they fit into certain categories.

The categories that the students created were limited to the number of groups that were to be formed for the duration of the laboratory session meetings. This turned out to be six groups of three to four students. The overarching themes ended up being: science; political infrastructure; resource management and sustainability; protection; sociological concerns; and, communications and navigation.

During each meeting of the laboratory session, students would hand in their laboratory exercises as directed by the Activities Manual, and then address their topics by further discussion and break up into questions, posed for all to see, and share out.

Thus, the students spend all laboratory periods focused on their semester-long problem or project, culminating in a written paper, with oral presentations during the last laboratory session meeting.

At the last laboratory session meeting, all students were asked to sign consent-to-publish forms, so that their written papers could be used in an edited volume. The results of which are presented following this introductory chapter.

0.2   THE SCIENCE OF ASTROBIOLOGY

It is emphasized that his course is a science course for non-science majors, and was taught with the cognizance that this was likely to be the last science course that most of the students would take.

In lecture, and in laboratory sessions, it was emphasized that astrobiology is the scientific study of the origin, development, distribution and search for life in the universe. This led to an acronym, originally developed by the lead instructor when this course was first taught over a decade ago, that is ODDS. This acronym was promoted as a memory aid device for students to recall the key goals for astrobiology.

Astrobiology is a multidisciplinary subject which touches upon the traditional sciences of physics, chemistry, geology, biology and astronomy. The highlights of the astrobiology concepts included the following classical components:

- Newton's Laws of Motion
- Newton's Universal Law of Gravitation
- Kepler's Law of Planetary Motion
- Einstein's Special Theory of Relativity
- $E = m c^2$
- Einstein's General Theory of Relativity
- 1st and 2nd Laws of Thermodynamics
- Energy Concepts [potential, kinetic, etc.]
- Electricity and electronics
- Computers and their physical limits
- Conservation of Momentum
- Light and spectroscopy
- Distances to the Stars

- Temperatures of the Stars
- Composition of the Stars
- Abundance of Chemical Elements
- Life Cycle Development of the Stars
- Death of Stars
- The Interstellar Medium
- Galaxies, clusters of galaxies, superclusters of galaxies, and the large scale structure of the universe
- Big Bang Cosmology and Inflation
- The Nature of Life
- The Evolution of Life
- The Evolution of Man
- The Search for Life in the Solar System
- The Drake Equation
- The Search for Extraterrestrial Life
- The Fermi Paradox

0.3   THE LEARNING MANAGEMENT SYSTEM

The astrobiology course utilized computer technology in the form of an interactive web based syllabus which included links to all lectures and questions posed utilizing the iClicker personal response system. The laboratory component relied upon a learning management system (LMS) which is utilized on the campus of George Mason University (GMU) that is the Blackboard LMS.

The figure below is a screen capture of the LMS, depicting the six groups from the laboratory session, and their main focus. Each group was responsible for its material, including the peer-reviewed references discovered by the groups, and the questions that they were addressing, all during the semester.

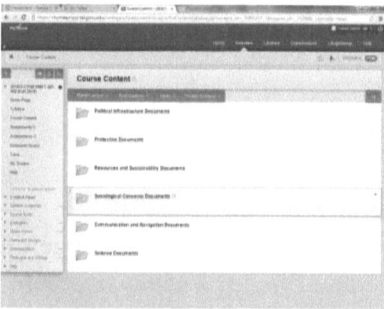

The figure below is another screen capture from the Blackboard LMS, this time depicting the different sections for the draft report and discussions of their topics. The wiki feature of the Blackboard LMS was utilized for allowing all students to participate, within the confines of their groups and their group emphasis.

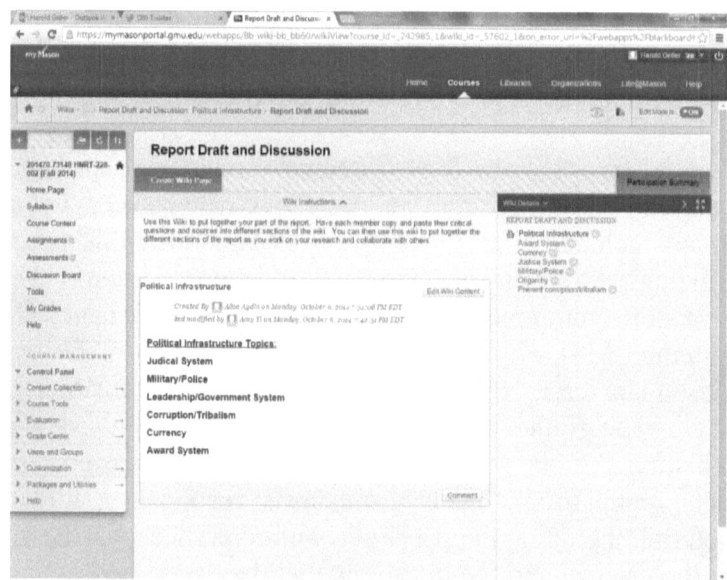

Ultimately, the final reports were deposited on the Blackboard LMS and retrieved by the GTA and the instructor for grading purposes and incorporation into this edited volume.

0.4 COURSE STRUCTURE

The astrobiology course was conducted as part of the Honors College track for non-science students. Within this track students must take two semesters of a lab-based science. Originally, the Honors College required a multidisciplinary science course, using an integrated science textbook by Trefil and Hazen for the first semester. Later, requirements were relaxed and any one semester laboratory based science course within the College of Science was allowed for credit towards the first semester. The second semester had students take one of the specially designed laboratory based science courses designed by faculty from different departments within the College of Science. Astrobiology was a topic chosen for the second semester science course for Honors College students because of its interdisciplinary nature and its focus in the media today.

The astrobiology course is taught in what has now become known as the "flipped" manner. However, the so-called "flipped" lecture session is not new by any means. The editor of this volume was in a history class over 40 years ago which utilized this style of teaching. Within this schema, students are asked to read the material on their own, before coming to lecture. The instructor then reviews the material, and queries the students continuously, to uncover if the students understood the material they were to have read, and guide the students into better understanding. It should be noted that within the Honors College at GMU, the census is held to 20 to 25 students.

Unlike the situation forty years ago, today such pedagogy is supplemented with computer technology, in this case, a personal response system (PRS), specifically iClicker. The instructor uses the PRS as a formative assessment tool, asking questions and seeing responses in real time.

After the review of the chapter's material, with the help of the PRS, the students were separated in groups of three to four. Each group was to choose a leader, recorder and one or two other participants. The role of the students was changed in round robin fashion, after each lecture meeting period.

All groups were assigned three to five thought provoking questions, which were chosen so that students could demonstrate what they had learned from the chapter material. The responses were graded for their completeness in sentences, critical thinking skills, and scientific reasoning. It was emphasized to all students that scientific reasoning was not just good critical thinking, but also the application of physical laws, which they had been exposed to during earlier sessions.

Group responses were turned in before students left for their next class periods. The group responses were graded and returned to students during the next class meeting, where students could share out their responses and guided by the instructor as to the validity of their responses.

0.5   PROBLEM BASED LEARNING LABORATORY

As alluded above, the laboratory sessions were also "flipped" in a sense. The students were assigned activities from the Activity Manual for Life in the Universe. The results of the activities were turned into the GTA at the start of each laboratory session. The laboratory session then continued by having the laboratory groups address overarching questions or problems, and then share out their questions that they felt were needed to address these overarching issues.

The ultimate goal was to address the multigenerational starship mission to another star. The

GTA would guide the discussions of the questions, and ascertain that all students from all groups were participating.

The students first decided that the laboratory instructor provide them with a definitive destination. After a couple of sessions, the laboratory instructor chose Kepler 186f as the destination.

Kepler 186 is a star studied by the Kepler mission using the so-called transit method. That is, the star is studied with the aid of a photometer, and small changes in the light emanating from the star allow scientists to determine if there may be any planets orbiting. As it turns out, Kepler 186, which about 500 light years distant from the Earth, has a number of planets orbiting its star. The planet known as Kepler 186f was determined to orbit the star Kepler 186, at a distance that allows for the presence of liquid water on its surface.

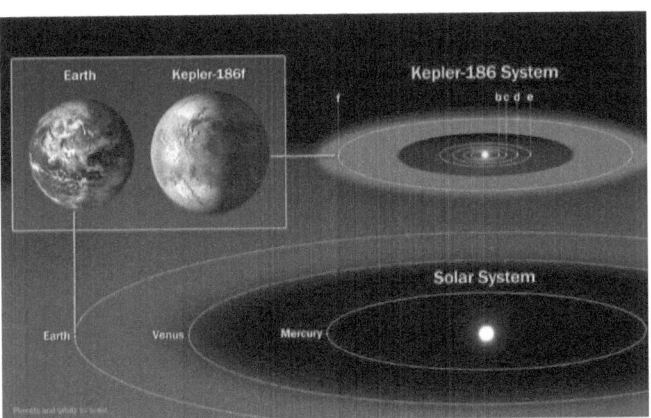

Image Credit: NASA Ames/SETI Institute/JPL CalTech

Kepler 186 is not a star like our own Sun. It is lower in mass, and glows with a reddish hue. Due to its lower mass and surface temperature, Kepler 186f must be closer to its star than our Earth is to our Sun. In fact, Kepler 186f is so close to its host star that it only takes about 130 days to orbit Kepler 186.

Kepler 186f itself is also about forty percent larger than the Earth. Nonetheless, it is located in what is called the habitable zone of Kepler 186, which again, only means that the temperature on the surface of Kepler 186f is suitable for supporting liquid water.

While the students chose to demand a final destination, they chose to overlook the means of propulsion that would take the starship to its destination. To supplement the work done by the enrolled Honors College students in areas they chose not to highlight, the course instructor was able to work with three high school students to fill in some of the overlooked areas.

There were three high school students who participated in this effort. Two focused on the propulsion systems for interstellar starship design, and one on the science experiments that would be the focus of the scientists on board the star ship.

After each of the laboratory groups had completed a time period wherein which they came up with questions to address under their overarching theme, all groups shared out their discussions, and they were guided by the GTA in better addressing the questions with critical thinking and scientific reasoning.

The questions from the sessions and some discussion were posted on the LMS wiki pages by the student groups.

The grades for the laboratory sessions were based upon the graded activities from the Activity Manual; the wiki input from the various question sessions; and, the final laboratory report.

0.6 GROUP GOALS

The focus of the student groups participating in the laboratory section was outlined with the help of the LMS.

That is, students posted numerous questions that they felt would be appropriate to address, and these questions were shared with their respective groups and the other students in the remaining groups.

The initial questions that the groups discussed included those addressing science, politics, protection, sustainability, sociology, communications and navigation; and, the questions and categories as defined by the groups are outlined here:

Science Questions
- Is Kepler 186f likely to be habitable and what are conditions likely to be on the planet?
- What do astronauts need?
- How are we going to collect and report data back to Earth?
- What is our goal—colonization or bring back data?

Political Infrastructure
- Judicial System
- Military/Police
- Leadership/Government System
- Corruption/Tribalism
- Currency
- Award System

Spaceship Protection
- How do we protect from interstellar debris?
- How do we protect from cosmic radiation?

Life Sustainability
- How to make food sustainable.
- How do we create a suitable climate and how to renew resources like Oxygen, CO_2, temperature, pressure, etc.?

Sociological and Governance
- What types of jobs are necessary to be self-sustaining?
- What type of education system/ instructional

materials will we implement?

Communications and Navigation

- How to navigate?
- Should we send a smaller ship ahead that we could store extra fuel on or a small probe
- Communication is needed but what information is important and priority?
    - What information will be useful to send back to earth, i.e. star maps, route inefficiencies, etc.?

0.7   CONCLUSION

The following chapters, exclusive of those regarding the interstellar medium, relativity, and the search for life in the universe, have been written by the participating students, as edited by the editor of this volume. The resultant papers derived from the laboratory session discussions and subsequent efforts at home, are presented as a set of sample outcomes from such a problem based laboratory experience. Educators using this approach will of course have their own students' views of the important questions regarding a multigenerational starship design. We hope that such outcomes will be educational for both the students and the instructors. We also hope that other educators who take this approach publish their students' papers and share out with a much wider audience. May the stars be in our future. Ad astra.

0.8   REFERENCES

De Graaff, E. and Kolmos, A. (2003). Characteristics of problem-based learning.*International Journal of Engineering Education. 19*(5), 657-662.

Frazier, W. M., & Sterling, D. R. (2008). Problem-based learning for science understanding. *Academic*

Exchange Quarterly, 12(2), 111.

Hmelo-Silver, C.E. (2004). Problem-based learning: What and how do students learn? *Educational Psychology Review, 16*(3).

Krajcik, J., Blumenfeld, P. C., Marx, R. W., Bass, K. M., Fredricks, J., & Soloway, E. (1998). Inquiry in project-based science classrooms: Initial attempts by middle school students. *Journal of the Learning Sciences, 7*(3-4), 313-350.

Matkins, J. J., Sterling, D. R., McDonnough, J. T., & Frazier, W. M. (2012). Constructing the science methods course as a shared instructional product. A paper for the Annual meeting of the National Association of Research in Science Teaching, Indianapolis, IN.

Sterling, D. R., & Frazier, W. M. ASTE Conference 2010 Sacramento, CA.

Chapter 1: An Investigation of Propulsion
By Sowmya Ranga

1.1 INTRODUCTION

Interstellar space travel remains a human goal waiting to be achieved. Although many ideas for such travel have been proposed, in the light of our current technology, interstellar travel is still preposterous. Many reasons exist for a manned interstellar mission. One reason is to investigate space and enhance our understanding of the science beyond the boundaries of our Solar System. Another reason is curiosity and the innate human nature to explore. To achieve interstellar space travel we must consider our current technology. Ideally, an interstellar mission to a nearby single star in the galaxy would be valuable to study and explore that star's system and planets, if it were to have any. One of the nearest stars to our Solar System, approximately 5.9 light years away, is Barnard's Star. Barnard's star is suspected of possessing at least two planets in orbit around it, with one having similar size and shape characteristics to Jupiter (Pande, Guven, & Velidi, 2011). Imagine if such a mission went forward in the near future. It would revolutionize science, especially the fields of astronomy and astrobiology, even if no forms of life were discovered. However, Barnard's Star is still significantly far for any successful mission with current travel methods. If you could imagine a spacecraft that could travel at the speed of light, the fastest attainable speed possible, it would still take about 11.8 years for a round trip to Barnard's Star, as measured by those on Earth.

Another possible destination, despite being of less interest to extraterrestrial life enthusiasts, is the multiple star system of Alpha Centauri. Of this system, Proxima Centauri would be the closer star to travel to, at a distance of 4.3 light years. Achieving space travel to Proxima

Centauri would set the stage for other interstellar travel. While succeeding in a manned mission out of the solar system, ground-breaking discoveries about the medium of space beyond the heliopause would be observed first-hand by an experienced crew. With hopes of making such a mission a reality, innovative new ideas and techniques are continually explored.

Due to the distance to the stars and limited velocity of the speed of light, only a multi-generational spacecraft appears feasible now, or in the future. To propel a multi-generational spacecraft on an interstellar journey requires serious consideration of alternative propulsion systems. Many propulsion methods exist, both in theory and in practice. However, while each has its own advantages, each also has drawbacks. When discussing spacecraft propulsion, the three functions of propulsion must be addressed. The propulsion system must be able to transfer the spacecraft from orbit about the Earth, into trajectories for interplanetary or interstellar travel. The propulsion system must also be able to orient the spacecraft once in orbit and when in interplanetary or interstellar space (Braeunig, 2008).

For one perspective of a manned interstellar space mission, let us consider the following. To understand the energy requirement and other limiting factors, let us consider the target distance of Proxima Centauri, the nearest star located 4.2 light years from Earth. For those less familiar with the distance of a light year, consider that 4.2 light years is equivalent to a distance of 39,734,219,300,000 kilometers. For this thought experiment let us consider a manned spacecraft accelerating half way to the target and then decelerating the second half of the trip, to control its landing or stopping upon reaching the target and to avoid overshooting the targeted star system. Ideally, to simulate an Earth-like experience for the astronauts aboard, the acceleration

16

would be equal to gravity (*g*), or roughly 9.8 meters per second per second. If the astronauts are positioned on board in parallel with the direction of travel, then the acceleration of *g* would allow them to experience Earth-like gravity. The maximum velocity that could be achieved by traveling to Proxima Centauri with this acceleration is about 95% of the speed of light. Using this calculation, and applying Einstein's relativistic time dilation formulae, it would take about 5.8 years, with respect to the time observed on earth, to reach the destination.

A problem arises though when the logistics of this mission are considered. The energy required to fuel such a mission, in this case assuming nuclear fusion for propulsion, is prohibitive. If the spacecraft mass were a total of 2 million kilograms, using a fuel conversion rate of 0.008 kilograms per meter per meter for hydrogen into helium fusion, the energy needed for this journey can be calculated to be about 780,000 exajoules. The problem arises when this number is taken into perspective with respect to the current world's energy consumption. In 2008, 474 exajoules of energy were consumed globally. Thus this mission becomes unrealistic with the required energy (Geffen, 2012).

Many other problems hindering such a space mission to Alpha Centauri become clear with this example scenario. There are many factors required to develop an ideal set of criteria for a manned space mission. In addition to acquiring the necessary specific impulses and attaining the desired velocities, an acceptable acceleration must also be determined. In the above given scenario, an acceleration of one *g* was used to recreate Earth-like conditions aboard the spacecraft. It is possible, though, for humans to adjust to slight fluctuations of *g*, sometimes up to 2*g*. Nonetheless, these are still just a portion of the details in constructing such a mission (Pande et al., 2011).

1.2   CHEMICAL PROPULSION

Chemical propulsion has been the most effective and successful method for propulsion in the past, and is still the most widely utilized.   To produce a thrust using chemical propulsion, an effective propellant with a reasonable specific impulse is required. The propellant consists of a fuel and an oxidizer, that is, an agent that releases oxygen to combine with the fuel. This approach burns the mixture when combined, ejecting an exhaust gas, producing thrust. The greater the speed of the exhaust gas ejection, the better the specific impulse.   The specific impulse, measured in seconds, is characteristic of the efficiency of the propellant used. Obtaining a good specific impulse is dependent upon various conditions, including a high combustion temperature within the spacecraft for burning the fuel, exhaust gas with little molecular weight, and also quality operating conditions and design of the spacecraft (Braeunig, 2008).

Solid, liquid, and hybrid propellants are used in rocket propulsion. There are three distinct types of liquid propellant used in chemical propulsion: cryogenic, petroleum, and hypergolic. Petroleum propellant is crude oil and a mixture of complex hydrocarbons, which is a compound made of hydrogen and carbon. Usually in rocket propulsion, this is kerosene, or refined kerosene (RP-1), used in combination with liquid oxygen as the oxidizer.

Cryogenic propellant is a liquefied gas, specifically liquid hydrogen (LH_2), which must be stored in low temperatures to sustain its liquid form. It is also used in combination with a liquid oxygen oxidizer. The maintenance of this low-density, low-temperature propellant, requires high-volume, heavy tanks to contain the substance. This becomes disadvantageous as that would contribute significantly to the payload. However, these drawbacks can be neglected when the high efficiency of cryogenic combination of liquid hydrogen and liquid oxygen propellant is taken into account. This propellant fuel, liquid hydrogen, is used mainly for powering the upper stages of spacecraft launches for its high specific impulse delivery, which is about 40% higher than other propellants.

Hypergolic propellants are highly toxic and do not require combustion to activate. The spontaneous ignition of the fuel and oxidizer, triggered by immediate contact with each other, make such propellants extremely dangerous to handle. Hypergolic propellants are liquid in form at standard temperature and pressure and can therefore, be stored easily in tanks. Because these propellants do not require an ignition source and react merely upon contact, they are ideal for spacecraft maneuvering systems with their capacity for easily starting and restarting. Commonly found hypergolic fuels include unsymmetrical dimethyl hydrazine (UDMH), monomethyl hydrazine (MMH), and hydrazine combined with typically a nitrogen tetroxide (N_2O_4) or nitric acid (HNO_3) oxidizer. Hydrazine is also used as a mono-propellant which acts in the presence of a catalyst that decomposes liquid hydrazine into hot gas, producing a specific impulse of about 230 or 240 seconds. Another popularly used fuel can be a combination of the previously named fuels, such as, Aerozine 50, a 50-50 mixture of UDMH and hydrazine.

Solid propellant motors, arguably the simplest of rocket designs, are typically used in the final stages of spacecraft launches. Unlike liquid propellants engines, solid propellant motors cannot be stopped once ignited and must burn until the propellant is completely exhausted. Another shortcoming of solid propellants is their high densities which contribute significantly to the payload weight. However, they have higher performance compared to liquid propellants, and similar to hybrid propellants. The fuel and oxidizer in solid propellants are mixtures of solid compounds and are stored in a, typically steel, casing of the solid propellant motor. Because solid propellants burn from the center outwards, the shape of the center channel is indicative of the rate and pattern of the burning, allowing some control of the thrust produced by the hot gases expelled by the burning. Of the solid propellants, there are two main types, both of which are dense, stable, and easily storable at normal temperatures, that is, homogeneous and composite.

A simple base homogeneous propellant, consisting of a single compound, has the capacity of oxidation and reduction. Nitrocellulose is typically used for this process. Alternatively, double base homogeneous propellants consist of three compounds typically: nitrocellulose, nitroglycerine, and a plasticizer. Composite propellants are made up mixtures using crystallized or finely round mineral salt for an oxidizer. Although the fuel is aluminum, generally ammonium perchlorate constitutes from 60% to 90% of the propellant mass. The final propellant, sometimes consisting of additional compounds such as a catalyst for rapid burning, is bound together by a polymeric binder with an overall consistency of a hard rubber eraser.

Hybrid propellant engines are an intermediary, between solid and liquid propellant engines, consisting of one solid and one liquid as each of the fuels and oxidizers. The liquid, usually the oxidizer, is infused with the solid, usually the fuel, in the solid storage chamber, enacting combustion. Hybrid propellant combustion can therefore be controlled to some extent as it can be stopped, restarted, or moderated. Unfortunately, this type of propellant is rarely used because of its inability to produce large thrusts (Braeunig, 2008).

Chemical propulsion derives its energy through the chemical reactions whose products are expelled as exhaust gas producing the thrust. Therefore, the chemical propulsion systems are limited by the reaction energies available.

1.3   ELECTRIC PROPULSION

Electrical propulsion varies from chemical propulsion in that the system uses electric energy to accelerate the exhaust gas to a much higher velocity than compared to chemical systems. A commonly used element in an electric system of propulsion is xenon. Because xenon remains in a gaseous phase at standard temperature and pressure, it does not require extra energy for any further conversions. The ion thruster in such a system of propulsion ionizes the gas and then electrically accelerates the gas. Xenon, when accelerated to a speed of about 30 kilometers per second, can create a thrust force strong enough to push the spacecraft in the opposite direction (Ward, 2000a).

Electric propulsion requires energy from an outside power source. This may be in the form of nuclear energy, solar radiation, or batteries. Power supplies become a major constraint for electric thrusters as they can significantly limit the thrust levels of the spacecraft. Although the payload of the spacecraft would be much lighter in an electric propulsion system, the power source must supply a significant power input. We can utilize the power and thrust force equations for electric propulsion (*EQ-5*, *EQ-6*, *EQ-7*) to elucidate the limitations of electric energy propulsion.

Consider a 1000 Watt power supply, about the power of an industrial vacuum cleaner. If the conversion efficiency is assumed to be 1.0 for simplicity, and the effective exhaust velocity is 30,000 meters per second, typically the velocity needed for a xenon electric propulsion system, then the resulting thrust force would be a mere 0.067 Newtons. Obviously the thrust levels of an electric propulsion system are significantly lower than that of chemical propulsion systems (Erichsen, 2006).

Solar energy is considered a good source of energy here on Earth. However, solar panels would not be efficient in an interstellar space mission as the source would not be sufficient even at the limits of the Solar System. Therefore, the method of solar sail propulsion cannot be considered feasible for any interstellar missions. Solar sail spacecraft operate by reflecting solar radiation or beams projected from Earth to produce acceleration (Ward, 2000b). This method of propulsion has had success with the Japanese IKAROS spacecraft launched in 2010 (Malik, 2010).

There are three basic types of electric propulsion systems: electrothermal, electromagnetic, and electrostatic. Each of these utilizes different methods to accelerate the propellant. Typically, more than one single method is implemented in a practical propulsion system.

Within an electrothermal propulsion system there are three methods used to accelerate the propellant by electrically heating the substance. Resistojets are one method in which a gas propellant is passed through an electric heater, allowing the gas particles to expand to produce the thrust. Usually this method is used in combination with a chemical propulsion system to further heat the propellant and create a more optimal specific impulse with a greater exhaust velocity.

Arcjets are another type of electrothermal propulsion systems. This method is used to attain exhaust velocities higher than 10,000 meters per second by passing the propellant through an electric arc and heating the flow to exceed 10,000 Kelvin. Inductively heated or radiatively heated systems are the third sub-type. These systems heat the propellant flow through some form of electrode-less discharge. An oscillating electromagnetic field is utilized in heating the free electron component of the propellant gas.

There are two types of electromagnetic propulsion systems that have been developed. Both systems use orthogonal electric and magnetic fields to accelerate the propellant. One is the pulsed plasma thruster and the second is the magnetoplasmadynamic thruster. The pulsed plasma thruster method, typically used in low power propulsion systems, less than 30 watts, utilizes electric energy stored in a capacitor to create a pulsed arc discharge over the surface of a quantity of propellant. Thermal flux, particle bombardment, and surface reactions combine to strip and ionize a small amount of the solid propellant substance. By acting on the ions moving in the electric field, a self-induced magnetic field causes the discharge to create a Lorentz force to accelerate the plasma.

Magnetoplasmadynamic thrusters work in a similar manner. These thrusters pass a current radially outwards through a neutral plasma from a center cathode to an outer ring-shaped anode. A Lorentz body force is created again and acts on moving ions in the discharge current to accelerate the plasma. This mechanism requires a high power supply and therefore is not currently feasible to serve as the primary propulsion method in an interplanetary mission, much less an interstellar mission. Pulsed plasma thrusters would also not be feasible as they would not produce sufficient thrust for such a mission.

There are three types of propulsion systems that are referred to as electrostatic propulsion systems. All three systems utilize electric fields to accelerate ionized propellant. The field effect electrostatic propulsion system ionizes the propellant substance by inducing a strong electric field. This method can produce an exhaust velocity of 10,000 meters per second. Systems with colloidal thrusters electrically accelerate charged, sub-micron diameter droplets of a liquid propellant of a conducting, non-metallic substance. Finally, ion thruster propulsion systems, discussed earlier, are used commonly with xenon propellant. In an ionization chamber, the propellant gas is ionized and accelerated near the exit by a double-grid structure and the electrons are extracted from the chamber by an anode. These electrons are pumped by the power supply to the cathode to be neutralized again after exiting the chamber. It must be noted that only ion engines have had success in spacecraft missions. Colloidal thrusters are still a new, developing technology and field effect electrostatic propulsion systems produce insufficient thrust to act as primary propulsion systems ("Electric spacecraft propulsion," 2004).

1.4   NUCLEAR PROPULSION

Although fusion propulsion systems are still a developing technology, they have potential as an effective primary propulsion system. Because the fusion of hydrogen into helium requires and produces colossal energy and heat, this method of propulsion has not been very successful to date, largely due to the lack of any known material that could possibly facilitate the process, that is, contain the hot plasma. However, plasma is a good conductor of electricity and, therefore, the plasma created by fusion may possibly be held, guided, and accelerated using magnetic fields. If the tremendous amount of energy capable of a nuclear fusion reaction can be harnessed in propulsion, then substantial advancements in propulsion technology would revolutionize space travel (Bonsor, 2001).

To successfully achieve interstellar travel within a realistic time frame, specific impulses of at least 30,000 seconds must be attained, which seems possible with a nuclear fusion propulsion system. This long-lasting fuel system would allow round trips to Proxima Centauri and other nearby stars to be possible within one lifetime (Pande et al., 2011).

Fusion propulsion would revolutionize space travel if successful. A fusion propelled spacecraft could attain a specific impulse about 300 times greater than chemical rocket propulsion systems. A fusion propulsion system may acquire a specific impulse of 130,000 seconds, considerably greater than a mere 450 seconds produced by a conventional chemical rocket engine. Nuclear fusion requires hydrogen as the propellant for the fusion to take place. This is an advantage in the fact that hydrogen is the most abundant chemical element throughout the universe, let alone the Solar System. A hydrogen fusion fueled propulsion system could replenish its source of hydrogen by extracting some from the atmosphere of an extrasolar planet. This process of replenishment would have to be further studied to ensure spacecraft safety and to determine a method for that extraction.

The efficiency of using fusion propulsion is notably beneficial when considering the travel time of a mission. With such a large specific impulse, a fusion-powered rocket would be able to travel longer distances in shorter time spans, in effect exhausting less propellant. Manned interplanetary travel would become much more feasible and convenient (Bonsor, 2001).

Nuclear fission is another technology that is being advocated in advancing propulsion technology, in a manner similar to nuclear fusion. Nuclear fission is typically proposed within a nuclear fusion system or vice versa. Essentially, nuclear fission propulsion is a system in which the neutron bombardment of, usually, uranium or plutonium nuclei, releases energy used to propel the spacecraft (Rudo, 2003). The released energy, or radiation, is used to heat the propellant and expel it through the nozzle, ideally a magnetic nozzle which can withstand the heat produced by the reactions. The initial liquid state of propellant is heated to a gas and its flow is accelerated by an enriched process of nuclear fission with the radioactive elements. Normally, because Uranium is a naturally occurring radioactive element, it is generally utilized in the form of Uranium-235, for these nuclear reactions. In the case of Uranium, the bombardment of the uranium would release tremendous amounts of energy which would be used as thermal energy to propel the gas of the propellant. This is the typical process of a nuclear thermal propulsion system ("Nuclear Reactors for Space," 2013).

One proposed method of nuclear propulsion for space travel is the use of a Nuclear Thermal Reactor (NTR). The advantage of utilizing thermal propulsion in a nuclear propulsion system is that the extreme temperatures reached in the combustion chamber of the rocket are almost directly proportional to the exhaust velocities achieved by those rockets. Therefore, the higher the temperatures in the combustion chamber, the higher the exhaust velocities. This technique would require a simple fission reactor to generate the necessary heat for the process in the NTR's core. The propellant, namely hydrogen for its compatibility and beneficial characteristics, is accelerated through the reactor's core. The heat causes the propellant to reach expeditious speeds as the flow of it is compressed and expelled through a durable nozzle. With the temperature of the flow reaching up to 3000 Kelvin, high exhaust velocities are easily attainable. Even so, specific impulses only surpass the 500 second mark, still creating insufficient amounts for extensive space travel. However, this method of propulsion would greatly benefit interplanetary space travel, for example, to Mars (Pande et al., 2011).

Nuclear electric systems are another form of nuclear propulsion systems. In this case the electric propulsion mechanism is powered by nuclear reactors to propel a spacecraft. The nuclear reactors serve as a heat source for the electric ionization process. By ionizing the propellant, typically hydrogen or xenon for this electric method, the superconducting magnetic cells can heat up to exceptionally high temperatures, namely millions of degrees Celsius, and thus accelerate the ionized propellant to approximately 30 kilometers per second and eject it to produce thrusts. Nuclear electric systems are especially beneficial as they can propel spacecraft already in space ("Nuclear Reactors for Space," 2013).

Nuclear energy as a source of propulsion is a strongly advocated research topic that may hold great potential for future propulsion systems. Although there are many safety considerations, and the need for the energy supply to induce the process, the results predicted by this propulsion system would ultimately be worth it.

1.5   ANTIMATTER PROPULSION

If there were any possibly stronger propulsion systems other than the ones discussed above, that may be a matter-antimatter propulsion system. The phenomenal energy produced by the collision of matter and antimatter would definitely be able to power a spacecraft into interstellar space. As a result of the interaction between the matter and antimatter, an explosion occurs in which both the masses of the particles are converted to energy, a 100 percent energy conversion. These explosions emit pure radiation that travels at the speed of light. This energy is almost 10 billion times the energy that chemical reactions would release. Moreover, these matter-antimatter reactions would be 1000 times more powerful than the nuclear fission energy in power plants and 300 times more powerful than nuclear fusion energy. With this powerful energy source and ideal speed of energy expulsion, space travel would be incredibly simple and realistic within a single lifetime on earth.

Antimatter is simply the exact opposite of matter. There exist positrons, antiprotons, and anti-atoms, which were first created, possibly by the Big Bang. These antimatter particles are all the opposites of particles of normal matter. They contain the masses as their corresponding matter particles but are simply of an opposite charge. Positrons are positively charged electrons. Antiprotons are negatively charged protons. And anti-atoms are atoms with positrons and antiprotons.

The problem, however, is that there is obviously a lack of antimatter on earth. If antimatter existed on earth, then there would be constant collisions and explosions of antimatter and matter as they remained in contact and would continually release excessive energy and light. Clearly since that is not the case and earth thrives with matter, it may be possible that the Big Bang produced more matter than antimatter, resulting in the abundance of matter from the remains of the explosions of the original matter and antimatter. Some scientists, though, claim that there exists antimatter at the center of the galaxy. This would mean that antimatter does naturally exist in the universe and could be extracted if possible.

Currently, the European Organization for Nuclear Research, the CERN, creates anti-atoms in labs. The scientists create this antimatter through the use of high-energy particle colliders. The technology is essentially constructed of large tunnels lined with powerful supermagnets that propel atoms at speeds nearly equal to light speed. An atom accelerated through the tunnel will slam into a target and break down into particles. By force of the encompassing magnetic fields, some of the particles are separated into anti-particles.

A matter-antimatter engine would consist of three main parts: magnetic storage rings, a feed system, and a magnetic rocket nozzle thruster. The rings would hold the antimatter and keep them in motion around the magnetic rings to store them away from the matter until needed for energy. The feed system would enable the antimatter to be released when needed to collide with a target of matter and thus, release energy. Finally, the magnetic rocket nozzle thruster would be used to direct the energy created by the collision and expel it to produce thrust. With this system, approximately 10 grams of antiprotons would be necessary to fuel a manned spacecraft to Mars in only one month.

With technology present in the labs, as described above, about one to two picograms of antiprotons are produced in a year. The CERN researchers have successfully created nine anti-hydrogen atoms. However, these anti-atoms only lasted for 40 nanoseconds each, not even slightly enough to propel a spacecraft. In fact, a year's production of antiprotons at CERN would be enough to light a 100-watt electric bulb for a mere three seconds. Evidently, this technology would need to go a long way before becoming a realistic candidate for interstellar travel (Bonsor, 2000).

1.6   CONCLUSION

In reality, an interstellar mission is almost impossible with the current available technologies and resources. Though some of the propulsion methods presented above are very detailed and strong in theory, they are not realistic feasible due to natural limitations and lack of advanced technologies to bring to life the proposed ideas. Only some of the most developed systems of propulsion were explained. Each of them has their own drawbacks as well as considerable advantages. While chemical propulsion has been traditionally used for years and has proven to be successful for innumerable space missions, electric propulsion seems to be more efficient. Nonetheless, both these forms of energy stand no chance against nuclear or matter-antimatter energy. The latter two are undeniably the most powerful energy producing propulsion methods. To yield the most efficient and effective propulsion systems, two or more different propulsion theories could be combined. For example, as mentioned earlier, fusion and fission are strongly compatible propulsion mechanisms that would enhance the quality and power of a spacecraft when used together. Electric propulsion combined with fission or fusion would also be beneficial in other ways such as for propulsion needed after take-off during the travel. On the other hand, matter-antimatter propulsion systems, when eventually feasible, would revolutionize the field of space travel with their near 100 percent energy conversion rates, and practically attainable light speeds. Nonetheless, the energy and supply necessary to fuel the spacecraft is still a phenomenal amount hindering success of such propulsion systems with the present state of economy and science.

The benefits of interstellar travel are infinite. A mission into deep outer space would enable the crew to experience the universe beyond the realm of the solar system. It would provide opportunities to conduct science experiments and investigate the extent of the science believed and practiced on earth. Data from unmanned missions may have been very useful as well, but with pure machinery, there come many flaws and discrepancies. The technology may not be reliable. Moreover, a first-hand human experience would revitalize the field of astronomy with live observations from a human perspective.

To ensure a successful mission into interstellar space, many essential factors must be examined from the mechanics and engineering of the ship to the safety and health of the crew. Facing many difficulties in simply the propulsion aspect of the mission, the complexity of planning such a trip becomes apparent. Therefore, to make manned interstellar mission possible in the future, serious considerations for advancements must be initiated now. Through similar studies and investigations, the present limitations facing such a mission can be examined and overcome. Thus, with more research and study, there remains hope for sending humans beyond the solar system one day in the near future.

1.7 APPENDIX OF EQUATIONS

$$E = mc^2$$

EQ-1

Einstein's Relativity Equation
E - (J) - Energy
m - (kg) - mass
c - (m/s) - speed of light = 2.99792458 x 108 m/s

$$F = ma$$

EQ-2

Newton's Law
F - (N) - Force
m - (kg) - mass of rocket
a - (m/s^2) - acceleration

$$F = m\frac{\Delta v}{\Delta t} = v\frac{\Delta m}{\Delta t} = v_e \dot{m}$$

EQ-3

F - (N) - Force (thrust)
m - (kg) - mass of rocket
v - (m/s) - velocity
v_e - (m/s) - effective exhaust velocity
$\dot{m} = \Delta m/\Delta t$ - (kg/s) - mass flow rate

$$I_{sp} = \frac{V_e}{g}$$

EQ-4

Specific Impusle
I - (s) - (Specific) Impulse
V_e - (Ns) - exhaust velocity
g - (m/s^2) - gravity (acceleration with respect to earth)

$$E_{jet} = \frac{1}{2}m_p v^2_{\,e}$$

EQ-5

Kinetic Energy of Ejected Matter
E_{jet} - (Ws) - energy
m_p - (kg) - mass of propellant
V_e - (Ns) - exhaust velocity

$$P_{jet} = \frac{dE_{jet}}{dt} = \frac{1}{2}\dot{m}v^2_{\,e} = F\frac{v_e}{2}$$

EQ-6

Power of the Jet
P_{jet} -(W) - power
E_{jet} - (Ws) - energy
m_p - (kg) - mass of propellant
$\dot{m} = \Delta m/\Delta t$ - (kg/s) - mass flow rate
V_e - (Ns) - exhaust velocity
F - (N) - Force (thrust)

$$P = \frac{P_{jet}}{\eta} = \frac{1}{2\eta}\dot{m}v^2_e = F\frac{v_e}{2\eta}$$

EQ-7

Power Input
P - (W) - power input
η - (constant) - power conversion efficiency
P_{jet} -(W) - power
E_{jet} - (Ws) - energy
m_p - (kg) - mass of propellant
$\dot{m} = \Delta m/\Delta t$ - (kg/s) - mass flow rate
V_e - (Ns) - exhaust velocity
F - (N) - Force (thrust)

$$T = 2\sqrt{\frac{k}{a}\left[\left(\frac{d}{2k} + 1\right)^2 - 1\right]} \qquad \text{where} \qquad k = \frac{c^2}{a}$$

EQ-8

T - (s) - travel time as observed on earth
c - (m/s) - the speed of light = 2.99792458 x 108 m/s
a - (m/s^2) - the constant acceleration/deceleration
d - (m) - the distance to the destination v_{max} - (m/s) - the maximum velocity obtained at the halfway point in the journey

$$V_{max} = \frac{c}{\sqrt{1 + \frac{k}{a\left(\frac{T}{2}\right)^2}}}$$ where $k = \frac{c^2}{a}$

EQ-9

v_{max} - (m/s) - the maximum velocity obtained at the halfway point in the journey
c - (m/s) - the speed of light = 2.99792458 x 108 m/s
a - (m/s^2) - the constant acceleration/deceleration
T - (s) - travel time as observed on earth

$$E_{kg} = 2c^2 \left(\frac{1}{\sqrt{1 - \left(\frac{V_{max}}{c}\right)^2}} - 1 \right)$$

EQ-10

E_{kg} - (J/kg) - the energy needed per kilogram of payload to make the journey
v_{max} - (m/s) - the maximum velocity obtained at the halfway point in the journey
c -(m/s) - the speed of light = 2.99792458 x 108 m/s

$$E = rm_{fuel}c^2$$

EQ-11

E - (J) - Energy- Assuming that 100% of the fuel is used for the propulsion of the spacecraft
r - a constant; the rate at which the fuel is converted to energy
m_{fuel} - (kg) - mass of the fuel
c - (m/s) - speed of light = 2.99792458 x 108 m/s

$$U_{gi} + K_i = U_{gf} + K_f \qquad\qquad \textbf{\textit{EQ-12}}$$

U_{gi} - (J) - the initial gravitational potential energy
K_i - (J) - the initial kinetic energy
U_{gf} - (J) - the final gravitational potential energy
K_f - (J) - the final kinetic energy

$$U_g = mgh$$
$$\textbf{\textit{EQ-13}}$$

U_g - (J) - the gravitational potential energy
m - (kg) - the mass of the object
g - (m/s^2) - the gravitational acceleration
h - (m) - the height of the object from an arbitrarily defined base level of zero height

$$K = \frac{1}{2}mv^2 \qquad\qquad \textbf{\textit{EQ-14}}$$

K - (J) - the kinetic energy
m - (kg) - the mass of the object
v - (m/s) - velocity of the object in motion

1.8   REFERENCES

Bonsor, K. (2000, November 29). How antimatter spacecraft will work. Retrieved January 29, 2014, from How Stuff Works website: http://science.howstuffworks.com/antimatter1.htm

Bonsor, K. (2001, March 12). How fusion propulsion will work. Retrieved January 29, 2014, from How Stuff Works website: http://science.howstuffworks.com/fusion-propulsion1.htm

Braeunig, R. A. (Ed.). (2008). Rocket propellants. Retrieved January 29, 2014, from http://www.braeunig.us/space/propel.htm

Electric spacecraft propulsion. (2004, June 15). Retrieved January 29, 2014, from http://sci.esa.int/smart-1/34201-electric-spacecraft-propulsion/?fbodylongid=1537

Erichsen, P. (2006, September). *Chapter 1: Introduction to spacecraft propulsion* [pdf]. Retrieved from http://fred.unis.no/AGF218/Handout_Erichsen_Propulsion.pdf

Geffen, N. (2012). Interstellar space travel is really difficult! Retrieved January 29, 2014, from http://spacetravel.nathangeffen.webfactional.com/spacedifficulties.php

Malik, T. (2010, July 12). Solar sail passes big test in deep space. Retrieved January 29, 2014, from Space.com website: http://www.space.com/8748-solar-sail-passes-big-test-deep-space.html

Nuclear reactors for space. (2013, December). Retrieved January 29, 2014, from World Nuclear Association website: http://www.world-nuclear.org/info/Non-Power-Nuclear-Applications/Transport/Nuclear-Reactors-for-Space/

Pande, S., Guven, U., & Velidi, G. (2011). Interstellar spaceflight using nuclear propulsion and advanced techniques. Retrieved January 29, 2014, from Academia.edu website: http://www.academia.edu/1102968/Interstellar_Spaceflight_Using_Nuclear_Propulsion_And_Advanced_Techniques

Rudo, B. (2003, March 5). Nuclear propulsion and what it means to space exploration. Retrieved January 29, 2014, from Red Colony website: http://www.redcolony.com/art.php?id=0303050#null

Ward, D. (2000a). Electric (ion) propulsion. Retrieved January 29, 2014, from Spark website: http://eo.ucar.edu/staff/dward/sao/fit/electric.htm

Ward, D. (2000b). Solar sails. Retrieved January 29, 2014, from Spark website: http://eo.ucar.edu/staff/dward/sao/fit/sails.htm

Chapter 2: On-the-Trail Science
By Sonia Thakur

2.1 INTRODUCTION

Needless to say, an interstellar journey of any kind is no simple undertaking. It would be helpful to start planning as soon as possible, which starts now. There is much in space yet to be discovered. Some predict that even with all of our current advancements in technology, we have only discovered approximately four percent of the observable universe (Das, Mena, Palomares-Ruiz, & Pascoli, 2013).

The focus of this chapter is to find the most feasible experiments to conduct during an interstellar journey if we do so sometime in the future. This chapter will examine the pros and cons of each experiment in order to decide which ones to prioritize. The goal of this chapter is to explore what is needed to be investigated in the future when humans are in a journey into interstellar space. This adds another purpose to the journey itself, other than meandering through the emptiness of space. The experiments would test if what we believe about certain science matters are true and give further insight into the nature of space. Obviously, experimentation is simpler if it is conducted in space instead of in vitro. There would be fewer hindrances and a higher yield of results.

The various components that are examined for each experiment includes: the instrumentation that would be used, and the hypotheses that would be examined in the experiments conducted. This chapter is organized in sections by the distance from Earth, with a section for experiments that would be conducted at all distances.

2.2   Science at the Heliopause

After the multigenerational starship leaves the Solar System, it will travel through the heliopause, the space between the Sun and the interstellar medium, where the Sun's solar wind ends its journey.

Our local galactic environment is more complicated than originally thought. An example for what could be done in the vicinity of the heliopause is what Voyager 1 and the Interstellar Boundary Explorer (IBEX) have done when they were at the heliopause. This experiment is about finding more about the interaction of the heliosphere with the very local interstellar medium. More needs to be known about the overall structure of the heliosphere. The spaceship would first travel into the very local interstellar medium, or VLISM.

There is a proposed band of atomic hydrogen emission that roughly surrounds the Sun with varying intensities. The spaceship can travel into an area with higher hydrogen emission intensity in this band which is on the ecliptic longitude of the Sun of about 221 degrees and ecliptic latitude of about 39 degrees.

At least a few measurements should be made. The shocked, solar-wind flow speed should be measured first. Then energetic neutral atoms (ENA) could be measured. While IBEX was measuring from the Earth's orbit, the spaceship would be able to have more direct and close measurements.

If the solar-wind were not flowing in the direction of the band, then the solar wind would not likely be the cause for the band of emission. In situ measurements of the ENAs could provide more of an idea of what is causing the band of emission and discover more about the overall structure of the heliosphere. When the spaceship travels near the outside of the heliopause, it could conduct more measurements to check if there are any other forces that could be affecting the structure of the heliosphere structure by conducting other experiments such as plasma-wave and magnetometer experiments. This experiment will also try to verify the idea that most of the energy density in the plasma is in a non-thermal constituent which may broaden to higher energies.

There are a few instruments involved in such experiments. One instrument would have to be able to measure a wide range of energies from the ENAs, from about 200 to 6 keV. The neutral-atom imagers would need to be more sensitive and weigh less than 25 kg. Another instrument would have to measure the shocked, solar-wind flow speed in order to provide more evidence that the wind speed is not causing the band of atomic hydrogen emission.

This experiment would help discover more about our local galaxy and would show easier trajectory paths for future spacecraft from Earth, with the possibility of more assistance from the gravity of Jupiter. The people on the spaceship would still be able to communicate with the people on Earth, so the results of this experiment may encourage future missions (McNutt Jr., Gruntman, Krimigis, Roelof, & Wimmer-Schweingruber, 2011).

2.3   SCIENCE BEYOND THE HELIOPAUSE

After the heliopause, is the interstellar medium, the space between star systems. Here, there are dust, gas, and cosmic rays. Much is unknown about this space. The ship will spend most of its time in the various interstellar mediums.

Not much is known about dark matter, or whether it even exists. Some tests that suggest the existence of dark matter are the Big Bang nucleosynthesis, the cosmic microwave background, and the large scale structure power spectra (Ciarcelluti & Wallemacq, 2014) The best way, however, to prove the existence of dark matter is to encounter it somehow, which the spaceship probably will do so, given enough time. Heavier dark matter particles of a few GeV can be captured by the stars and produce high energy neutrinos by decays from Standard Model particles. Dark matter mass can helped be determined by neutrino telescopes that detect these high energy neutrinos (Das, Mena, Palomares-Ruiz, & Pascoli, 2013). It is assumed that about ten GeV dark matter particles annihilate into leptons and that about 50 GeV particles annihilate to quarks. A possible astrophysical explanation for the signal is millisecond pulsars (Hooper & Slatyer, 2013).

There are ways to detect dark matter as well. It is beneficial to search for dark matter out in space, not just because the location of most dark matter is unknown, but also because measuring near Earth, an irreducible neutrino background comes in the way of observing. The detectors would be able to constrain properties such as mass scattering cross section and spin. In this way, the dark matter density of the area local to the spaceship at the time would be determined, especially with our dark matter halo (Baudis, 2012). The difference between light WIMP dark matter and hidden sector dark matter should also be determined. Light WIMP dark matter uses contact interactions through a single type of particle. On the other hand, hidden sector dark matter has many components and interacts through a mediator with no mass. Using the Germanium recoil spectrum, one can find that that the hidden sector dark matter has a sharply rising nuclear recoil spectrum. This is how the two possibilities can be differentiated (Foot, 2013).

Dark matter is mostly thought of now as something that is cold and almost completely collisionless. This is why it is sometimes referred to as cold dark matter. But because of the complexity and vastness of the visible world, a more complex dark sector is a possibility. This leads to a concept called Double-Disk Dark Matter

(DDDM), where dark matter interacts with itself and loses enough energy to obtain dynamics similar to the baryonic matter, where there are sides to obtain balance. In this model, the DDDM can cool enough to form disks within galaxies. Using observational signatures, finding that the density at the center as well as the plane of the galaxy could imply that there are DDDM (Fan, Katz, Randall, & Reece, 2013).

There is also the idea that dark matter could be related to mirror matter as well. Mirror matter is the counterpart of ordinary matter. Mirror matter is a way of restoring the symmetry of the laws of nature, which does not apply to dark matter at the moment. Dark matter could be partially or fully composed of mirror matter. When simulations were run, it was found that a consistent amount of mirror dark matter fit well along with the cold dark matter. How much mirror matter may be there, if at all, is unknown. According to the models, pure mirror matter, mirror matter mixed with cold dark matter, and pure dark matter fit with the data from the cosmic microwave background and the large scale structure power spectra, so studying those would not be as helpful anymore. The existence of mirror matter however, can be observed through gravitational effects. If the spaceship senses mirror matter, it could somehow affect the gravity near the mirror matter that some effect would be able to be observed (Ciarcelluti & Wallemacq, 2014). Another way of evidencing mirror matter is through dwarf galaxies. There were about half of dwarf galaxies observed orbiting around M31. If dark matter is collisionless then these dwarf galaxies should not have any dark matter. However, there are other observations leading to the idea that dwarf satellite galaxies are full of dark matter. This could imply that mirror matter is involved (Foot & Silagadze, 2013).

The spaceship would be likely encounter asteroids on its journey. The shape of various asteroids can be determined by radar and light curve observations (Nolan et al., 2013).Formulas can be used to predict the number of asteroids nearby that can bear useful ores. Near Earth, the number of asteroids nearby that may contain platinum is estimated to be ten. Over nine thousand contain water, but many of them are relatively small (Elvis, 2013). While traveling through space, the spaceship should be able to detect if approaching asteroids contain useful materials. A thermophysical model could be used, which takes into account the asteroid's orbit, spin, shape, and heat diffusion (Matter, Delbo, Carry, & Ligori, 2013).

2.4   EXOPLANETARY SCIENCE

The nearest star to our Sun is Alpha Centauri, so that will be the exoplanetary system to be studied. A possible planet that the ship could land on later on for a visit is Alpha Centauri Bb, although its existence has not been confirmed yet.

2.5   INTERSTELLAR SCIENCE

There are other experiments that would best be conducted throughout the trip as the spaceship goes through the various phases of space. Throughout the journey, the ship should be constantly measuring some features in space, such as temperature.

2.5.1   COSMIC DUST SCIENCE

There are various experiments that could be done with cosmic dust. The cosmic dust could be observed from outside the spaceship, or samples could be taken inside.

The amount of dust thought to be in our Solar System is actually overestimated. This is because the side walls of the dust detectors in the Ulysses and the Galileo instruments. These side walls take up about 27 percent of the space inside these instruments. Since the impacts on these side walls affect the time of the target ionization signal, the particle fluxes and velocity are overestimated. This means that the amount and extent of the cosmic dust in our own solar system is unknown. A simple beginning experiment on this journey would be to measure the amount of cosmic dust encountered while covering a specific area of the solar system. It is important to estimate the amount of cosmic dust within our own solar system and to check if the amounts are overestimated (Willis, Burchell, Ahrens, Krüger, & Grün, 2005).

Studying the way particles aggregate in space could lead to more knowledge about the way the planetesimals formed in our early Solar System. The way the cosmic dust scatters light should be observed. As more particles aggregate, the instruments would measure how the polarization evolves. The instrument should be able to observe the cosmic dust from various angles of the light (Lasue & Levasseur-Regourd, 2007).

The instruments will be similar to the Active Cosmic Dust Collector used in the Stardust mission. It would use instrumentation that involves aerogel and aluminum foil collectors to pick up some cosmic dust for sampling. The polarization tool used would be the Light Scattering Unit from the Interactions in Cosmic and Atmospheric Particle System facility. The trajectories for the cosmic dust particles would be found by allowing charged particles to fly through a positron sensitive electrode system and measuring induced electric signals. So far, cosmic dust particles as small as 0.4 micrometers can be measured, but the sensitivity for size of particles should be increased for this experiment (Grün et al., 2012)

Silicon and olivine grain has also been found to be more useful materials for collecting cosmic dust. Analytic scanning electron microscopy can be used to characterize the various grains of the cosmic dust. A light-gas-gun and accelerators could also be used to acquire samples of the comic dust. Once acquired, however, the dust needs to be extracted and refined. This extraction can be done by using a 266 nm UV laser microdissection system that would recover the olivine grains. The dust can be cleaned up by focusing an ion beam milling to remove excess aerogel that could still be on the surface of the grains (Graham et al., 2004).

These experiments and observations would try to find the origin of many of the cosmic dust particles in order to understand the diversity of the interstellar medium. This would also help understand the kinds of celestial objects that may have been near the region of interstellar medium that the spaceship would be in at various times.

2.5.2 COSMIC RAY SCIENCE

Cosmic rays are another important area of study. They propagate all throughout space and would be able to be detected at all stages of the mission. Cosmic rays can usually be measured by different instruments such as ionization chamber, meson telescopes, and neutron monitors (Berezinsky, 2014).

Cosmic rays even affect our atmosphere. Background neutrons are produced because of the interaction of cosmic rays with the atmosphere. Such cosmic ray produced background secondary neutrons are being recorded by the Lead-Free Gulmarg Neutron Monitor, which provides more evidence to the extent of cosmic rays (Darzi, Ishtiaq, Mir, Mufti, & Shah, 2014).

The spaceship would be able to see the effects of cosmic rays in the heliosphere as well. Only cosmic rays in the energy range of ten to the powers of seven and ten eV are normally sensitive to the dynamics to the heliosphere, so an instrument would need to be able to focus in on those energies while the spaceship is travelling through the heliosphere. The heliosphere becomes transparent for higher energies, and particles usually get locked in the solar wind for lower energies. Anomalous cosmic rays, or ones that have unexpected composition rather than spectral shape, should be specifically investigated when studying the movement and acceleration of energetic particles through the heliosphere (Moraal, 2014).

The study of cosmic rays continues as the spaceship moves through the interstellar medium. The acceleration from the sources of the cosmic rays and the propagation through the interstellar space are the two processes that shape the spectra of the cosmic rays. The rays then diffuse in magnetic fields. The previous part should be focused on in order to help find out where the cosmic rays are coming from (Ptuskin, 2012). There is also a feature called the cosmic ankle that should be investigated while the spaceship makes the journey outwards into the interstellar medium. This cosmic ankle has normally been seen as a feature caused by the transition from galactic to extragalactic cosmic rays. This causes a dip in the differential spectrum of the cosmic rays. Although the dip shape has been confirmed, the transition from galactic to extragalactic cosmic rays should occur at a lower energy. By measuring the energies and the spectrums while the spaceship is travelling, it might be able to discover if this cosmic ankle does occur at a lower energy. One more small and continuous experiment with cosmic rays is learning more about the general flux and frequency of extragalactic cosmic rays. For relatively low energies, a negligible amount of particles can enter our galaxy from any distance greater than 100 Mpc. This may be due to particle destruction in other galaxies along the way. Once the spaceship reaches other galaxies, it can observe more about the farther away cosmic rays (Berezinsky, 2014).

Cosmic rays could potentially affect the cosmic dust experiments that may be conducted on board the spaceship. When conducting low activity measurements with instruments such as Germanium gamma spectrometers, cosmic rays contribute a significant amount to the background signal which disturbs the experiments. The cosmic rays could penetrate through concrete and lead shielding. Cosmic rays should be carefully observed throughout the trip because they may affect many of the more precise experiments going on (Šolc, Kovář, & Dryák, 2014).

There is an interesting correlation between cosmic rays and dark matter. Cosmic rays could provide answers to questions about dark matter by providing energies higher those of accelerators. The interaction between cosmic rays and dark matter should be investigated (Grupen, 2014). When dark matter annihilates, the supposed idea is that some type of cosmic rays are the result. PAMELA and FERMI satellites have reported unexpected excess in the cosmic ray flux. These observations would be more meaningful when the spaceship was more out in space and had more data and time for collection of that data (Cirelli, 2011). A magnetic spectrometer could be used on the spaceship in order to search for these signatures of dark matter annihilating to charged cosmic rays (Carosi, Xiao, Fisher, Rybka, & Zhou, 2011). Antiprotons and antinuclei should be observed in the cosmic radiation because antiprotons are assumed to be produced by the interaction of cosmic ray protons with interstellar gas. A few antideuterons, but especially an antihelium, would help determine the existence and properties of dark matter (Chardonnet, Salati, & Taillet, 1999). If the spaceship does pass through the galactic center, it should be able to detect more cosmic rays from the center of the galaxy than from anywhere else if dark matter is composed of superheavy particles (Ananthaswamy, 2011).

Cosmic rays would be studied even when the spaceship approaches extrasolar systems. The signals are strongest from stars since they are the source of cosmic ray modulation. The instruments would likely observe electromagnetic observations at all wavelengths. The inner part of the heliosphere is the most dynamic region of it (Moraal, 2014). Cosmic rays could also be studied when looking at the potential planets while passing by. Lightning affects the formation of complex organic molecules that are likely to form life. There is a possibility that the increase in the flux of charged particles could cause the increase in the lightning rate. If the spaceship measures the flux of charged particles and observes the lightning on planets, then this may provide more evidence that there is a correlation. This observation would mean that cosmic rays could affect life, and this would helped be determined by observing the lightning on planets other than earth in order to get more variability (Atri & Melott, 2014).

2.5.3 CHEMICAL COMPOSITION

As the spaceship travels through various areas and phases of the galaxy and universe, it would also be able to measure the composition of its surrounding environment while travelling. If someone from Earth were to guess what the composition of an area light years away was, it would be difficult since that person would have to see through the composition of our local interstellar medium. These kinds of observations would at most only give the total composition in sight, not of what the composition is solely to each specific area. The only feasible way would be to travel through these phases. The composition could be measured by observing ions, with matrices, and by methods such as mass spectrometry (Danger et al. 2013).

2.6   CONCLUSION

The cosmic rays should be experimented on, because they affect so many other topics and encompass a variety of other problems. If there end up being any constraints on the number of experiments or cost or weight of the instruments, then cosmic rays should be studied because the most would be learned overall. The existence of dark matter is not yet known so it could be a waste of experimentation if it does not exist, especially with the idea of mirror matter. Asteroids could easily be deflected by a powerful spaceship. Although cosmic dust is one of the continuing experiments, it does not seem to have as much potential to reveal some discovery as studying cosmic rays does. The time spent studying our own heliosphere would be miniscule compared to the total amount of time spent on the other experiments since the spaceship would be travelling so far for so long.

The instruments need to be taken into consideration. The instruments need to be updated and fixed periodically. Since there are multiple experiments, and more to come, the total amount of instruments may contribute a significant amount of the weight of the spaceship. Voyager had ten instruments that weighed 104.32 kg in total, which was about 15 percent of the total payload mass ((McNutt Jr., Gruntman, Krimigis, Roelof, & Wimmer-Schweingruber, 2011). The more the instruments weigh, and the spaceship overall weighs, the harder it would be to propel the spaceship, especially when launching it out from Earth or from other exoplanets that may have higher gravities. Viewing angles, energy, and integration time all need to be considered thoroughly when trying to implement multiple instruments aboard the spaceship.

If all of these experiments were conducted, we would learn much more about space. Space is not best observed from the Earth. On the way, the ship may encounter objects and phenomena that we do not know of yet at all. Some experiments would need to be formed then to explore the said phenomena and objects. Although the people aboard the ship will be unlikely to communicate with the people still on Earth after a few years after the journey begins, the rest of space is waiting to be explored. This research will help plan for a potential future journey into interstellar space and indicate what could be considered beforehand. This is valuable because it would make progress in thinking about the future missions so it would one day be accomplished.

2.8 REFERENCES

Alexis Matter, Marco Delbo, Benoit Carry, Sebastiano Ligori, Evidence of a metal-rich surface for the Asteroid (16) Psyche from interferometric observations in the thermal infrared, Icarus, Volume 226, Issue 1, September–October 2013, Pages 419-427, ISSN 0019-1035, http://dx.doi.org/10.1016/j.icarus.2013.06.004.

Alison Gibbings, Massimiliano Vasile, Ian Watson, John-Mark Hopkins, David Burns, Experimental analysis of laser ablated plumes for asteroid deflection and exploitation, Acta Astronautica, Volume 90, Issue 1, September 2013, Pages 85-97, ISSN 0094-5765, http://dx.doi.org/10.1016/j.actaastro.2012.07.008.

Anil Ananthaswamy, Dark matter could produce ultra-high energy cosmic rays, New Scientist, Volume 195, Issue 2611, 7 July 2007, Page 8, ISSN 0262-4079, http://dx.doi.org/10.1016/S0262-4079(07)61673-X.

Chitta R. Das, Olga Mena, Sergio Palomares-Ruiz, Silvia Pascoli, Determining the dark matter mass with DeepCore, Physics Letters B, Volume 725, Issues 4–5, 1 October 2013, Pages 297-301, ISSN 0370-

2693,
http://dx.doi.org/10.1016/j.physletb.2013.07.006.

Claudio Bombardelli, Hodei Urrutxua, Mario Merino, Jesús Peláez, Eduardo Ahedo, The ion beam shepherd: A new concept for asteroid deflection, Acta Astronautica, Volume 90, Issue 1, September 2013, Pages 98-102, ISSN 0094-5765, http://dx.doi.org/10.1016/j.actaastro.2012.10.019.

Claus Grupen, Early developments: Particle physics aspects of cosmic rays, Astroparticle Physics, Volume 53, January 2014, Pages 86-90, ISSN 0927-6505, http://dx.doi.org/10.1016/j.astropartphys.2013.01.00 2.

D.K. Sharma, M.S. Khurana, Jagdish Rai, Ionospheric heating due to solar flares as measured by SROSS-C2 satellite, Advances in Space Research, Volume 48, Issue 1, 1 July 2011, Pages 12-18, ISSN 0273-1177, http://dx.doi.org/10.1016/j.asr.2011.02.007.

Dan Hooper, Tracy R. Slatyer, Two emission mechanisms in the Fermi Bubbles: A possible signal of annihilating dark matter, Physics of the Dark Universe, Volume 2, Issue 3, September 2013, Pages 118-138, ISSN 2212-6864, http://dx.doi.org/10.1016/j.dark.2013.06.003.

Debi Prasad Choudhary, Sanjay Gosain, Nat Gopalswamy, P.K. Manoharan, R. Chandra, W. Uddin, A.K. Srivastava, S. Yashiro, N.C. Joshi, P. Kayshap, V.C. Dwivedi, K. Mahalakshmi, E. Elamathi, Max Norris, A.K. Awasthi, R. Jain, Flux emergence, flux imbalance, magnetic free energy and solar flares, Advances in Space Research, Volume 52, Issue 8, 15 October 2013, Pages 1561-1566, ISSN 0273-1177, http://dx.doi.org/10.1016/j.asr.2013.07.009.

Dimitra Atri, Adrian L. Melott, Cosmic rays and terrestrial life: A brief review, Astroparticle Physics, Volume 53, January 2014, Pages 186-190, ISSN 0927-6505,

http://dx.doi.org/10.1016/j.astropartphys.2013.03.00
1.

E. Bellotti, C. Broggini, G. Di Carlo, M. Laubenstein, R. Menegazzo, Search for correlations between solar flares and decay rate of radioactive nuclei, Physics Letters B, Volume 720, Issues 1–3, 13 March 2013, Pages 116-119, ISSN 0370-2693, http://dx.doi.org/10.1016/j.physletb.2013.02.002.

E. Grün, Z. Sternovsky, M. Horanyi, V. Hoxie, S. Robertson, J. Xi, S. Auer, M. Landgraf, F. Postberg, M.C. Price, R. Srama, N.A. Starkey, J.K. Hillier, I.A. Franchi, P. Tsou, A. Westphal, Z. Gainsforth, Active Cosmic Dust Collector, Planetary and Space Science, Volume 60, Issue 1, January 2012, Pages 261-273, ISSN 0032-0633, http://dx.doi.org/10.1016/j.pss.2011.09.006.

G. Carosi, S. Xiao, P. Fisher, G. Rybka, F. Zhou, Searching for signatures of dark matter in the cosmic ray spectrum measured by AMS-01, Nuclear Physics B - Proceedings Supplements, Volume 221, December 2011, Page 335, ISSN 0920-5632, http://dx.doi.org/10.1016/j.nuclphysbps.2011.09.03
4.

G. Danger, F.-R. Orthous-Daunay, P. de Marcellus, P. Modica, V. Vuitton, F. Duvernay, L. Flandinet, L. Le Sergeant d'Hendecourt, R. Thissen, T. Chiavassa, Characterization of laboratory analogs of interstellar/cometary organic residues using very high resolution mass spectrometry, Geochimica et Cosmochimica Acta, Volume 118, 1 October 2013, Pages 184-201, ISSN 0016-7037, http://dx.doi.org/10.1016/j.gca.2013.05.015.

G.A. Graham, A.T. Kearsley, A.L. Butterworth, P.A. Bland, M.J. Burchell, D.S. McPhail, R. Chater, M.M. Grady, I.P. Wright, Extraction and microanalysis of cosmic dust captured during sample return missions: laboratory simulations, Advances in Space Research, Volume 34, Issue 11, 2004, Pages 2292-2298, ISSN 0273-1177, http://dx.doi.org/10.1016/j.asr.2003.07.066.

George Gloeckler, Johannes Geiss, Composition of the local interstellar medium as diagnosed with pickup ions, Advances in Space Research, Volume 34, Issue 1, 2004, Pages 53-60, ISSN 0273-1177, http://dx.doi.org/10.1016/j.asr.2003.02.054.

Hans J. Fahr, Roger Osterbart, Imprints of the interstellar medium on location, geometry and nature of the heliospheric shock, Advances in Space Research, Volume 13, Issue 6, June 1993, Pages 159-171, ISSN 0273-1177, http://dx.doi.org/10.1016/0273-1177(93)90405-Z.

Harm Moraal, Cosmic rays in the heliosphere: Observations, Astroparticle Physics, Volume 53, January 2014, Pages 175-185, ISSN 0927-6505, http://dx.doi.org/10.1016/j.astropartphys.2013.03.002.

Huirong Yan, A. Lazarian, Tracing magnetic fields with ground state alignment, Journal of Quantitative Spectroscopy and Radiative Transfer, Volume 113, Issue 12, August 2012, Pages 1409-1428, ISSN 0022-4073, http://dx.doi.org/10.1016/j.jqsrt.2012.03.027.

I.M. Podgorny, Yu.V. Balabin, A.I. Podgorny, E.V. Vashenyuk, Spectrum of solar flare protons, Journal of Atmospheric and Solar-Terrestrial Physics, Volume 72, Issue 13, August 2010, Pages 988-991, ISSN 1364-6826, http://dx.doi.org/10.1016/j.jastp.2010.05.010.

Ian A. Crawford, Project Icarus: A review of local interstellar medium properties of relevance for space missions to the nearest stars, Acta Astronautica, Volume 68, Issues 7–8, April–May

2011, Pages 691-699, ISSN 0094-5765, http://dx.doi.org/10.1016/j.actaastro.2010.10.016.

J. Lasue, A.C. Levasseur-Regourd, Cosmic dust optical properties: Numerical simulations and future laboratory measurements in microgravity, Advances in Space Research, Volume 39, Issue 3, 2007, Pages 345-350, ISSN 0273-1177, http://dx.doi.org/10.1016/j.asr.2005.05.010.

Jacques P. Vallée, Magnetic fields in the nearby Universe, as observed in solar and planetary realms, stars, and interstellar starforming nurseries, New Astronomy Reviews, Volume 55, Issues 3–4, May–June 2011, Pages 23-90, ISSN 1387-6473, http://dx.doi.org/10.1016/j.newar.2011.01.001.

Jaroslav Šolc, Petr Kovář, Pavel Dryák, MCNPX simulation of influence of cosmic rays on low-activity spectrometric measurements, Radiation Physics and Chemistry, Volume 95, February 2014, Pages 181-184, ISSN 0969-806X, http://dx.doi.org/10.1016/j.radphyschem.2012.12.034.

JiJi Fan, Andrey Katz, Lisa Randall, Matthew Reece, Double-Disk Dark Matter, Physics of the Dark Universe, Volume 2, Issue 3, September 2013, Pages 139-156, ISSN 2212-6864, http://dx.doi.org/10.1016/j.dark.2013.07.001.

Jonathan D Slavin, The radiation environment of the local interstellar medium, Advances in Space Research, Volume 34, Issue 1, 2004, Pages 35-40, ISSN 0273-1177, http://dx.doi.org/10.1016/j.asr.2003.01.033.

Juan José Curto, Luis R. Gaya-Piqué, Geoeffectiveness of solar flares in magnetic crochet (sfe) production: I—Dependence on their spectral nature and position on the solar disk, Journal of Atmospheric and Solar-Terrestrial Physics, Volume 71, Issues 17–18,

December 2009, Pages 1695-1704, ISSN 1364-6826, http://dx.doi.org/10.1016/j.jastp.2008.06.018.

Kunitomo Sakurai, A possible causal relation of the source composition of cosmic rays with the elemental depletion in the interstellar space, Nuclear Physics A, Volume 718, 5 May 2003, Pages 407-409, ISSN 0375-9474, http://dx.doi.org/10.1016/S0375-9474(03)00815-7.

L d'Hendecourt, E Dartois, Interstellar matrices: the chemical composition and evolution of interstellar ices as observed by ISO, Spectrochimica Acta Part A: Molecular and Biomolecular Spectroscopy, Volume 57, Issue 4, 15 March 2001, Pages 669-684, ISSN 1386-1425, http://dx.doi.org/10.1016/S1386-1425(00)00436-4.

L.I. Miroshnichenko, W.Q. Gan, Particle acceleration and gamma rays in solar flares: Recent observations and new modeling, Advances in Space Research, Volume 50, Issue 6, 15 September 2012, Pages 736-756, ISSN 0273-1177, http://dx.doi.org/10.1016/j.asr.2012.04.024.

Laura Baudis, Direct dark matter detection: The next decade, Physics of the Dark Universe, Volume 1, Issues 1–2, November 2012, Pages 94-108, ISSN 2212-6864, http://dx.doi.org/10.1016/j.dark.2012.10.006.

M. Cirelli, Dark Matter, Cosmic Rays and Neutrinos: status circa 2010, Nuclear Physics B - Proceedings Supplements, Volume 217, Issue 1, August 2011, Pages 237-242, ISSN 0920-5632, http://dx.doi.org/10.1016/j.nuclphysbps.2011.04.110.

M. Youssef, On the relation between the CMEs and the solar flares, NRIAG Journal of Astronomy and Geophysics, Volume 1, Issue 2, December 2012,

Pages 172-178, ISSN 2090-9977, http://dx.doi.org/10.1016/j.nrjag.2012.12.014.

M.A. Darzi, P.M. Ishtiaq, T.A. Mir, S. Mufti, G.N. Shah, Cosmic ray modulation studies with Lead-Free Gulmarg Neutron Monitor, Astroparticle Physics, Volume 54, February 2014, Pages 81-85, ISSN 0927-6505, http://dx.doi.org/10.1016/j.astropartphys.2013.11.01 0.

M.J. Willis, M.J. Burchell, T.J. Ahrens, H. Krüger, E. Grün, Decreased values of cosmic dust number density estimates in the Solar System, Icarus, Volume 176, Issue 2, August 2005, Pages 440-452, ISSN 0019-1035, http://dx.doi.org/10.1016/j.icarus.2005.02.018.

Martin Elvis, How many ore-bearing asteroids?, Planetary and Space Science, Available online 11 December 2013, ISSN 0032-0633, http://dx.doi.org/10.1016/j.pss.2013.11.008.

Michael C. Nolan, Christopher Magri, Ellen S. Howell, Lance A.M. Benner, Jon D. Giorgini, Carl W. Hergenrother, R. Scott Hudson, Dante S. Lauretta, Jean-Luc Margot, Steven J. Ostro, Daniel J. Scheeres, Shape model and surface properties of the OSIRIS-REx target Asteroid (101955) Bennu from radar and lightcurve observations, Icarus, Volume 226, Issue 1, September–October 2013, Pages 629-640, ISSN 0019-1035, http://dx.doi.org/10.1016/j.icarus.2013.05.028.

Mike Gruntman, Instrumentation for interstellar exploration, Advances in Space Research, Volume 34, Issue 1, 2004, Pages 204-212, ISSN 0273-1177, http://dx.doi.org/10.1016/j.asr.2003.04.064.

P. Chardonnet, P. Salati, R. Taillet, Antimatter cosmic rays, New Astronomy, Volume 4, Issue 4, July 1999,

Pages 275-282, ISSN 1384-1076, http://dx.doi.org/10.1016/S1384-1076(99)00033-0.

P.C Frisch, Why study interstellar matter very close to the Sun?, Advances in Space Research, Volume 34, Issue 1, 2004, Pages 20-26, ISSN 0273-1177, http://dx.doi.org/10.1016/j.asr.2003.02.068.

Paolo Ciarcelluti, Quentin Wallemacq, Is dark matter made of mirror matter? Evidence from cosmological data, Physics Letters B, Volume 729, 5 February 2014, Pages 62-66, ISSN 0370-2693, http://dx.doi.org/10.1016/j.physletb.2013.12.057.

R. Foot, Differentiating hidden sector dark matter from light WIMPs with Germanium detectors, Physics of the Dark Universe, Volume 2, Issue 2, June 2013, Pages 59-64, ISSN 2212-6864, http://dx.doi.org/10.1016/j.dark.2013.04.002.

R. Foot, Z.K. Silagadze, Thin disk of co-rotating dwarfs: A fingerprint of dissipative (mirror) dark matter?, Physics of the Dark Universe, Volume 2, Issue 3, September 2013, Pages 163-165, ISSN 2212-6864, http://dx.doi.org/10.1016/j.dark.2013.10.001.

R.T. James McAteer, Peter T. Gallagher, Paul A. Conlon, Turbulence, complexity, and solar flares, Advances in Space Research, Volume 45, Issue 9, 3 May 2010, Pages 1067-1074, ISSN 0273-1177, http://dx.doi.org/10.1016/j.asr.2009.08.026.

Ralph L. McNutt Jr., Mike Gruntman, Stamatios M. Krimigis, Edmond C. Roelof, Robert F. Wimmer-Schweingruber, Interstellar Probe: Impact of the Voyager and IBEX results on science and strategy, Acta Astronautica, Volume 69, Issues 9–10, November–December 2011, Pages 767-776, ISSN 0094-5765, http://dx.doi.org/10.1016/j.actaastro.2011.05.024.

Ralph L. McNutt Jr., Robert F. Wimmer-Schweingruber, the International Interstellar Probe Team, Enabling interstellar probe, Acta Astronautica, Volume 68, Issues 7–8, April–May 2011, Pages 790-801, ISSN 0094-5765, http://dx.doi.org/10.1016/j.actaastro.2010.07.005.

S. Ibadov, Space observations of comets during solar flares: A possible explanation for comet brightness outbursts, Advances in Space Research, Volume 49, Issue 3, 1 February 2012, Pages 467-470, ISSN 0273-1177, http://dx.doi.org/10.1016/j.asr.2011.11.001.

S.-I. Akasofu, D.N. Covey, Magnetic field configuration of the heliosphere in interstellar space, Planetary and Space Science, Volume 29, Issue 3, March 1981, Pages 313-316, ISSN 0032-0633, http://dx.doi.org/10.1016/0032-0633(81)90018-0.

Telemachos Ch. Mouschovias, Interstellar magnetic fields, Advances in Space Research, Volume 2, Issue 12, 1982, Pages 71-80, ISSN 0273-1177, http://dx.doi.org/10.1016/0273-1177(82)90290-3.

V. Berezinsky, Extragalactic cosmic rays and their signatures, Astroparticle Physics, Volume 53, January 2014, Pages 120-129, ISSN 0927-6505, http://dx.doi.org/10.1016/j.astropartphys.2013.04.001.

V.B. Baranov, Physical consequences from axisymmetric model of the solar wind interaction with supersonic interstellar gas flow, Advances in Space Research, Volume 16, Issue 9, 1995, Pages 307-319, ISSN 0273-1177, http://dx.doi.org/10.1016/0273-1177(95)00353-G.

Vladimir Ptuskin, Propagation of galactic cosmic rays, Astroparticle Physics, Volumes 39–40, December 2012, Pages 44-51, ISSN 0927-6505, http://dx.doi.org/10.1016/j.astropartphys.2011.11.004.

Chapter 3: Starship Astronautics
By Richard Oh

3.1 INTRODUCTION

Physicists and engineers are working to advance space propulsion systems to develop basic science knowledge. Ideas like Antimatter Initiated Microfusion (AIMStar), beamed propulsion, and ion engines have been studied, one of which has already been applied to real applications. To achieve interstellar travel however, current technology holds limitations that must be overcome. AIMStar propulsion systems need to address the containment issue with antimatter (as of now, scientists can contain 10^9 antiprotons for only 4-10 days, far less from the needed time of decades or even centuries for interstellar travel) (Lewis, Meyer, Smith, & Howe, 1999). For beamed propulsion, radiation beams need to be collimated and highly focused to direct momentum to the system's sail effectively. And the ion engine, though it has a high specific impulse, does not have practical applications for large payloads. In addition, costs for interstellar travel to the nearest star system, i.e. Alpha Centauri, are estimated to approach $122 trillion (Long, 2012).

Despite these obstacles, it is important to continue studying and advancing interstellar propulsion technology for the purpose of gathering invaluable data from exoplanets in nearby star systems. Within a 15 light year radius, a distance an interstellar probe can reach in under a century, there are 31 known stars that can be explored (Long, 2012). Collecting data from these stars and possibly their star systems by sending robot probes is a practical and achievable endeavor within the near future and a necessary step towards manned interstellar travel.

Space exploration outside the solar system may seem daunting considering that at the slow pace of current interstellar travel, it will take tens of millions of years to cross the Milky Way Galaxy. But eventually, there will be a dire need for humans to find refuge on other Earth-like planets due to overpopulation, and if not that the imminent threat of the Sun changing into a Red Giant. It is important to continue tackling the interstellar problem and running calculations to measure the limits for each type of propulsion system.

In this chapter, we analyze the behaviors of physical phenomena associated with rocket-propelled travel. By definition, a rocket is a propulsion system that carries all of its reaction mass, energy, and engines with it. A general computational platform that takes into consideration rocket mass, efficiency, exhaust speed, mass flow rate, and the relativistic effect was developed and studied for numerical calculations and results. All results were calculated with an underlying assumption that the acceleration felt by rocket passengers was constant.

Section 2 derives the general rocket equation from non-relativistic mechanics. Section 3 derives the general rocket equation that includes the relativistic effects, which behaves differently during the acceleration and deceleration – of equal magnitude – stage of the rocket travel. Section 4 presents the calculations and results. Section 5 describes the conclusions extracted from the results and applies them to practical applications to sending unmanned interstellar missions to nearby star systems and exoplanets.

3.2. NON-RELATIVISTIC ROCKET EQUATION AT CONSTANT ACCELERATION

The non-relativistic rocket equation relating mass and speed of the rocket was derived using the conservation of total linear momentum. A rocket with an initial mass M_0

ejects a portion of a reaction mass dm at a constant exhaust velocity relative to the rocket's reference frame w. Note that in order to maintain constant acceleration and deceleration throughout the rocket's travel, the amount of dm ejected from the rocket within a certain time must vary. Below is a figure illustrating variables of interest – where U is the rocket speed, M the rocket mass, w the exhaust speed relative to the rocket, dm the ejected reaction mass, and w_a the exhaust speed relative to an inertial reference frame. All vectors are parallel to one another.

Figure 1 Non-Relativistic Rocket Schematic Diagram

It is important to mention that in an ideal case, the rocket in the above figure is a point located at its center of mass as well as the ejected reaction mass. However, in a more realistic situation, dm is not a point, but a group of points each with its own mass and velocity collectively adding up to dm and w respectively. When scientists measure the exhaust speed w with an instrument, the measured speed is the average of all velocity vector of points. In an ideal case, for the group of points, all points are parallel to and do not collide with one another. As a result, in order to take into account of efficiency (in other words effective fuel utilization efficiency), w equals the multiplication of efficiency and ideal exhaust velocity (w = e*v). This relationship, however, assumes that there is a perfect conversion of fuel to energy. The more realistic and general equation should be w = e_{ff}*e*v where e_{ff} is the energy conversion efficiency. For convenience, let e_{ff}*e = e_f or rocket efficiency to yield a simple form of w = e_f*v. Below is an illustration to clarify. As depicted, in the Ideal

(Group of Points) category, the resultant velocity is larger than the measured exhaust speed w, thus a rocket efficiency variable should be multiplied to equal the measured exhaust speed w.

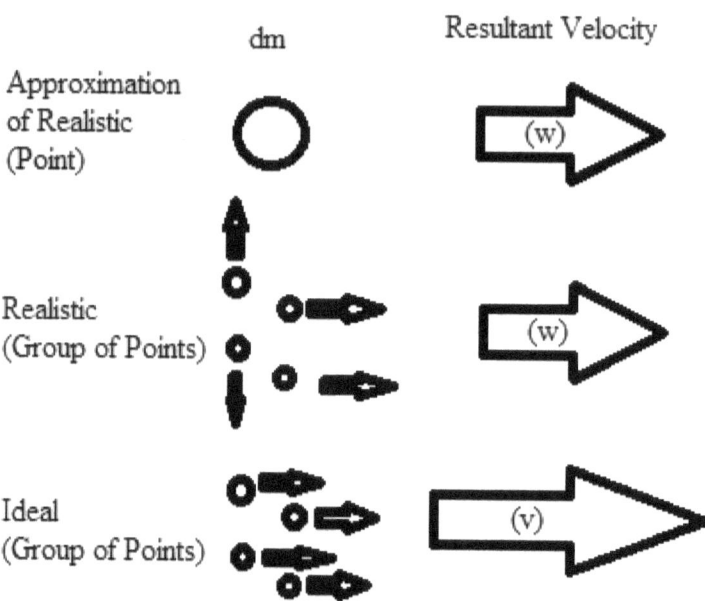

Figure 2 Illustration clarifying how the efficiency term was incorporated

From the law of linear momentum conservation, a relationship between the rocket's change in total linear momentum and the change due to the expulsion of the small fraction of the reaction mass dm

$$d(MU) = w_a * dm$$

where $dM = -dm$ by the conservation of mass and $w_a = w - U$ for the case of acceleration (Forward, 1995). By definition, acceleration is where U and w point at opposite directions, whereas during deceleration, U and w point in the same direction since deceleration is the decrease of the absolute value of speed. Then by plugging in these relationships with the conservation of linear momentum equation above, a new and simple relationship between rocket mass, rocket speed, and exhaust speed relative to the rocket is

$$\frac{dM}{M} = -\frac{dU}{w}.$$

After integrating both sides, the non-relativistic rocket equation is given by

$$M = M_0 e^{-\frac{U}{w}}.$$

Plugging in the relationships $w = e_f * v$ and $U = a * t + U_0$ – where a is the constant acceleration ($a > 0$), t the duration of travel time, and U_0 the initial velocity –, the non-relativistic rocket equation for acceleration is developed.

$$M(t) = M_0 e^{-\frac{at + U_0}{e_f v}}$$

This equation indicates that the mass of the rocket decreases as the rocket gains speed. For the deceleration non-relativistic rocket equation, the derivation can be reworked. However, when a rocket decelerates, it will consume the same amount of fuel as if the rocket was accelerating. There is no distinction between the rocket mass consumptions for the acceleration and deceleration cases, thus there is no need to develop a second non-relativistic rocket equation for the deceleration case, i.e. the rocket equation above is valid in both the acceleration and deceleration stages.

To derive the mass consumption rate in relation to the travel time t, first assume that the rocket accelerates from rest ($U_0 = 0$) for simplicity. The derivative of the non-relativistic rocket equation is

$$\frac{dM(t)}{dt} = \frac{M_0 a}{e_f v} e^{-\frac{at}{e_f v}}$$

where $\dfrac{M_0 a}{e_f v}$ is the initial mass flow rate $\dot{M}_0$. In other words, $\dot{M}_0$ is the rate at which mass is ejected from the rocket (Brewer, 1970). This equation describes how the mass consumption rate decreases as the travel time increases, which must occur due to the lower force needed to exert the same acceleration of a rocket with decreasing mass.

These derived equations were incorporated in the general computational platform to observe the behavior of a rocket that performed an acceleration stage during its first portion of travel and a deceleration stage for the latter portion. With these equations and basic kinematics equation, the relationship between travel time, distance covered, rocket acceleration, speed, mass, and mass consumption rate was developed.

3.3. RELATIVISTIC ROCKET EQUATION AT CONSTANT ACCELERATION IN THE ROCKET'S REFERENCE FRAME

For the derivation of the relativistic rocket equation, a similar process was used as described in Section 2, however the fundamental properties of time, mass, and acceleration vary with the rocket speed as seen by an outside observer. In order to incorporate the relativistic effect into the conservation of linear momentum derivation, the following relationships were used

$$\gamma = \left(1 - \frac{U^2}{c^2}\right)^{-1/2}$$

where γ is the Lorentz factor, U the rocket speed, and c light speed (= 2.998*10^8 m/s),

$$\Delta\tau = \frac{\Delta t}{\gamma}$$

where τ is the time elapsed inside the rocket's reference frame as observed by an outside observer in an inertial reference frame and t is the time elapsed in the inertial reference frame as experienced by the outside observer,

$$M = \gamma M$$

where M is the mass of the moving rocket and M is the rest mass of the rocket,

$$\alpha = \frac{a}{\gamma^3}$$

where α is the acceleration viewed by an inertial reference frame and a is the acceleration inside the rocket's reference frame (Knorr, 2010). Below is modified version of Figure 1 to illustrate the relativistic mass notations.

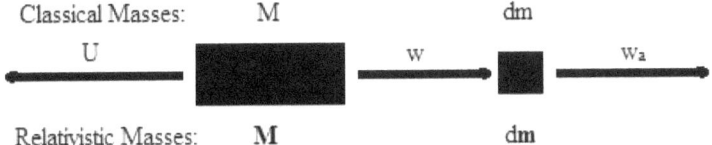

Figure 3 *Non-Relativistic and Relativistic Rocket Schematic Diagram during the Acceleration Stage*

3.3.1. RELATIVISTIC ROCKET EQUATION DURING THE ACCELERATION STAGE

Because nothing can travel faster than the speed of light, the physical phenomena during acceleration and deceleration stage behave differently especially as the rocket approaches to the speed of light. For the relativistic rocket equation relating the rest mass and speed of the rocket, the conservation of linear momentum gives

$$d(MU) = w_a * dm$$

where $d(Mc^2) = - c^2 * dm$ by the conservation of mass-energy
$$w_a = \frac{w - U}{1 - \frac{wU}{c^2}}$$
and for the case of acceleration (Forward, 1995) . After plugging in these equations, a relatively simple relationship between rest mass M, rocket speed U, and exhaust velocity relative to the rocket w is developed.

$$\frac{dM}{M} = - \frac{dU}{w\left(1 - \frac{U^2}{c^2}\right)}$$

After integration and simplifying, the relativistic rocket equation is

$$M(U) = M_i \left(\frac{1 + \frac{U}{c}}{1 - \frac{U}{c}}\right)^{-\frac{c}{2w}}.$$

While the w in the exponent can be rewritten as $e_f v$, the rocket speed $U(t)$ does not equate to the non-relativistic at (assuming that initial velocity is zero). Instead, the acceleration in the inertial reference frame needs to be integrated with respect to t as follows.

$$U(t) = \int_{t_0}^{t} \alpha(t)dt = a\int_{t_0}^{t} \frac{1}{\gamma^3}dt = a\int_{t_0}^{t} \left(1 - \frac{U(t)^2}{c^2}\right)^{3/2}dt$$

Differentiating with respect to t and isolating the variable $U(t)$

$$\frac{dU(t)}{dt} = a\left(1 - \frac{U^2}{c^2}\right)^{3/2} \rightarrow \frac{dU(t)}{\left(1 - \frac{U^2}{c^2}\right)^{3/2}} = adt,$$

then integrating both sides, the rocket speed as a function of t is

$$\int_0^U \frac{dU'}{\left(1 - \frac{U'^2}{c^2}\right)^{3/2}} = \int_0^t adt' \rightarrow U(t) = \left(\frac{1}{c^2} + \frac{1}{a^2 t^2}\right)^{-1/2}$$

where a is the constant acceleration inside the rocket's reference frame.

Plugging $U(t)$ and w into the relativistic rocket equation, the new relativistic rocket equation with respect to t is

$$M(t) = M_i \left[\frac{1 + \left(1 + \frac{c^2}{a^2 t^2}\right)^{-1/2}}{1 - \left(1 + \frac{c^2}{a^2 t^2}\right)^{-1/2}}\right]^{-\frac{c}{2e_f v}}.$$

When deriving the mass consumption rate, taking the derivative of $M(t)$ in respect to t is now meaningless because t depends on the speed of the rocket. Instead, $\frac{dM}{d\tau}$ is the meaningful mass consumption rate since to the passengers inside the rocket, they only care about the mass consumption rate in their own time scale, not the time scale in the inertial reference frame. In order to calculate the derivative, the time dilation relationship

$$d\tau = \frac{dt}{\gamma}$$

can be applied to $\frac{dM}{d\tau}$ to derive the relationship

$$\frac{dM}{d\tau} = \frac{dM(t)}{\left(\frac{dt}{\gamma}\right)} = \gamma \frac{dM(t)}{dt}.$$

With this, the mass consumption rate during the acceleration stage was determined to be

$$\frac{dM}{d\tau} = -\frac{M_0 a}{e_f v} \gamma \left(1 + \frac{a^2 t^2}{c^2}\right)^{-1/2} \left(\frac{1 + \left(1 + \frac{c^2}{a^2 t^2}\right)^{-1/2}}{1 - \left(1 + \frac{c^2}{a^2 t^2}\right)^{-1/2}}\right)^{-\frac{c}{2 e_f v}}$$

where $\dfrac{M_0 a}{e_f v}$ is the initial mass flow rate M_0.

3.3.2. RELATIVISTIC ROCKET EQUATION DURING THE DECELERATION STAGE

As shown in the figure below, for a rocket that began from rest, deceleration only occurs after acceleration because, by definition, deceleration is the decrease of the absolute value of speed. Thus, the initial rocket speed during deceleration U_1 is greater than zero, unlike during the acceleration stage when U_0 is assumed to be zero for simplicity. It is also important to note that due to the definition of deceleration, the derived rocket equation for deceleration can only be used for cases where U points in the same direction as w.

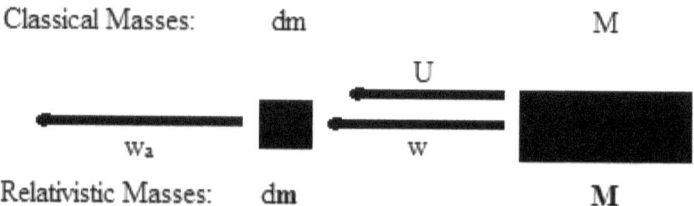

Figure 4 Non-Relativistic and Relativistic Rocket Schematic Diagram during the Deceleration Stage

70

Because both w and w_a during the deceleration stage are oppositely oriented from the w and w_a during the acceleration stage, they can be considered as $-w$ and $-w_a$, respectively, following the definitions in the acceleration case. After working through the conservation of linear momentum as described in Section 3a, except with the change in signage for w and w_a, a simple, now positive relationship appears as follows.

$$\frac{dM}{M} = \frac{dU}{e_f v \left(1 - \frac{U^2}{c^2}\right)}$$

Integrating both sides and evaluating gives

$$\int_{M_1}^{M} \frac{dM'}{M} = \ln\frac{M}{M_1} = \int_{U_1}^{U} \frac{dU'}{w\left(1 - \frac{U^2}{c^2}\right)} = \frac{c}{2e_f v} \ln\left[\left(\frac{1 + \frac{U}{c}}{1 - \frac{U}{c}}\right)\left(\frac{1 - \frac{U_1}{c}}{1 + \frac{U_1}{c}}\right)\right]$$

$$M(U) = M_1\left[\left(\frac{1 + \frac{U}{c}}{1 - \frac{U}{c}}\right)\left(\frac{1 - \frac{U_1}{c}}{1 + \frac{U_1}{c}}\right)\right]^{\frac{c}{2e_f v}}.$$

To write the relativistic rocket equation in terms of t, U during the deceleration stage needs to be derived. The derivation process for U is similar to the derivation shown in Section 3a, except the initial rocket speed U_1 cannot be assumed to be zero and a during the deceleration stage is negative, thus the following relationships are enforced.

$$U_1 > U(t) \geq 0$$
$$a < 0$$

The rocket speed during deceleration is described by

$$U(t) = U_1 + \int_{t_1}^{t} \frac{a}{\gamma^3} dt$$

Differentiating both sides in respect to t and isolating the variable of interest give the same equation form, the only difference being that a is negative.

$$\frac{dU(t)}{dt} = a\left(1 - \frac{U^2}{c^2}\right)^{3/2} \rightarrow \frac{dU(t)}{\left(1 - \frac{U^2}{c^2}\right)^{3/2}} = adt$$

Integrating both sides yields

$$\frac{1}{\left(\frac{1}{U^2} - \frac{1}{c^2}\right)^{1/2}} - \frac{1}{\left(\frac{1}{U_1^2} - \frac{1}{c^2}\right)^{1/2}} = a(t - t_1).$$

For simplicity, let

$$f(t) = a(t - t_1) + \frac{1}{\left(\frac{1}{U_1^2} - \frac{1}{c^2}\right)^{1/2}}.$$

Then solving for $U(t)$, a similar form as the $U(t)$ during acceleration is produced.

$$U(t) = \left(\frac{1}{c^2} + \frac{1}{f^2(t)}\right)^{-1/2}$$

As a result, the new relativistic rocket equation in respect to t is

$$M(t) = M_1 \left[\left(\frac{1 + \left(1 + \frac{c^2}{f^2(t)}\right)^{-1/2}}{1 - \left(1 + \frac{c^2}{f^2(t)}\right)^{-1/2}}\right)\left(\frac{1 - \frac{U_1}{c}}{1 + \frac{U_1}{c}}\right)\right]^{\frac{c}{2e_f v}}.$$

To determine the mass consumption rate in respect to τ, differentiate $M(t)$ in respect to t and multiply by the Lorentz factor to get

$$\frac{dM}{d\tau} = \frac{M_1 a}{e_f v}\gamma\left(1 + \frac{f^2(t)}{c^2}\right)^{-1/2}\left[\frac{1 + \left(1 + \frac{c^2}{a^2 t^2}\right)^{-1/2}}{1 - \left(1 + \frac{c^2}{a^2 t^2}\right)^{-1/2}}\right]\left(\frac{1 - \frac{U_1}{c}}{1 + \frac{U_1}{c}}\right)\right]^{\frac{c}{2e_f v}}.$$

where $\frac{M_0 a}{e_f v}$ is the initial mass flow rate M_1. It should be noted that for the deceleration stage, a is negative while the exhaust speed v is positive by definition.

The derived equations for both acceleration and deceleration stages were included in the general computational platform. With these equations, the relationship between travel time, distance covered, rocket acceleration, speed, mass, and mass consumption rate was developed.

3.4. Results and Analysis

Consider an ideal rocket with 100% efficiency, exhaust velocity of light speed, a strong thrust for constant acceleration of 9.81 m/s², and with all its mass (100,000 kg) as propellant. This extreme case would unveil the discrepancies between non-relativistic and relativistic predictions.

Traveling for 1500 days at constant acceleration of 9.81 m/s/s for the first half of the trip and constant deceleration of the same magnitude for the second half, the non-relativistic equations predict that the rocket reached a maximum speed of 2.12 times the speed of light (2.12*c) and traveled a distance of 4.35 light years (farther than the distance from Earth to Alpha Centauri star system), with a surplus of 1439.5 kg of reaction mass remaining by the end of the travel time. Below are the graphs depicting the non-relativistic calculations over the 1500 day period. Notice that Time t is the notation used for non-relativistic time scale t.

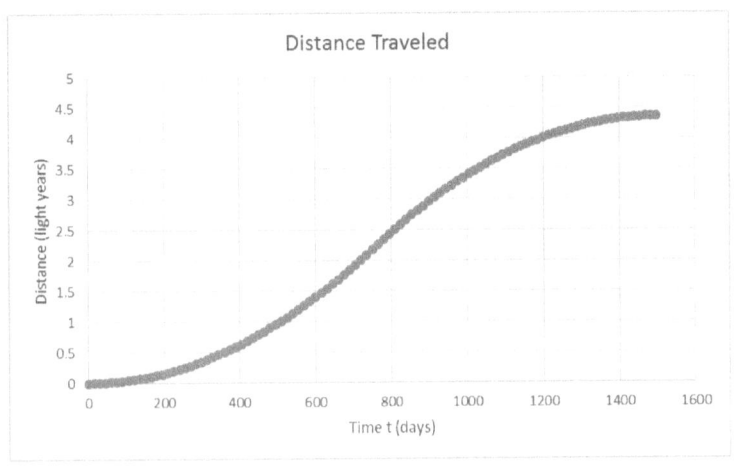

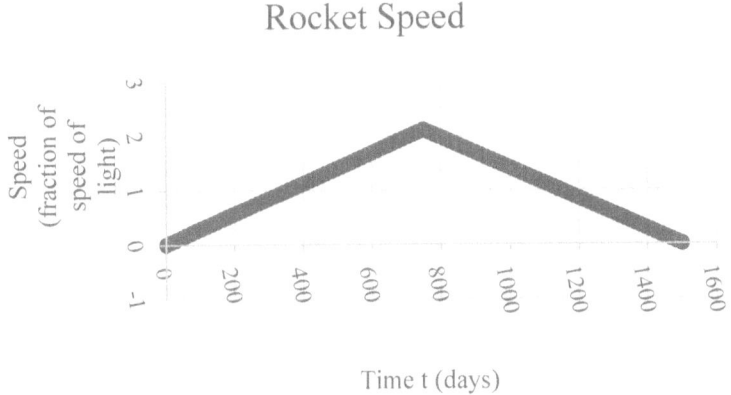

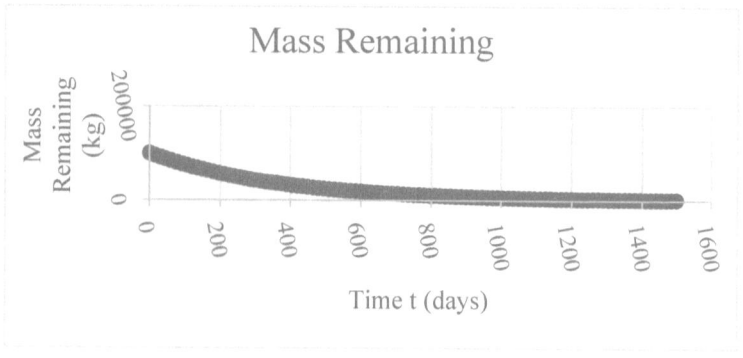

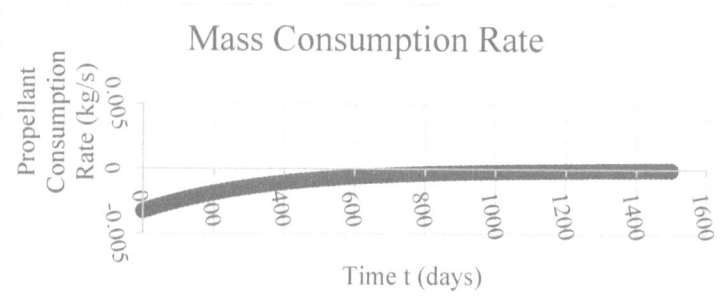

The distance and speed graphs illustrate the initial conditions that the rocket accelerates and decelerates for each half of its travel time. The mass remaining continuously drops exponentially, thus requiring less force to maintain constant acceleration or deceleration of the rocket. This corresponds with the exponential absolute value decrease of mass consumption rate.

The relativistic calculations differ significantly. Due to the upper speed limit (light speed), the concept of four dimensionality becomes important. Time is like a fluid, depending on the speed of the rocket. Though the projected time is aimed to 1500 days, this is in respect to the outside observer. According to the time scale inside the rocket, only 70.6% of projected time t elapsed. As illustrated in the next figure below, plotting time τ against time t, the relationship is not linear. Instead, it is curved with the slope approaching to zero as the rocket nears light speed at 750 days for time t. This can be shown more clearly in the Appendix. This indicates the time dilation effect, which is the precise reason why the total elapsed time τ is 70.6% of total elapsed time t.

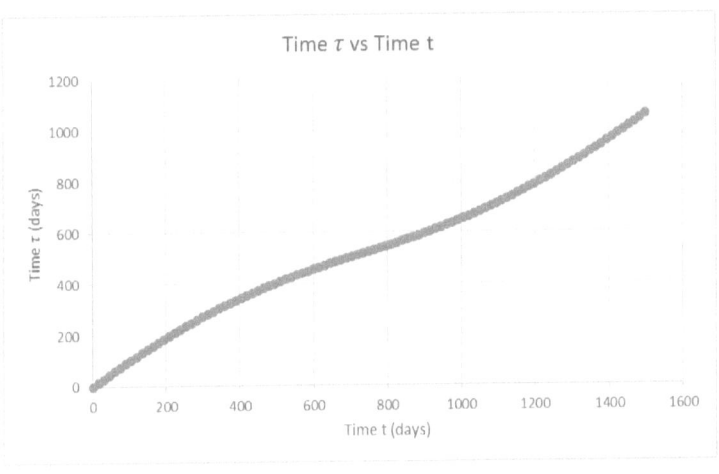

Another interesting finding is the acceleration of the rocket observed in the outside, inertial frame. Constant acceleration and deceleration of 9.81 m/s/s is important to rocket passengers because they should be in an environment with a similar gravitational pull as the Earth. As a result, the acceleration inside the rocket's reference frame is physically meaningful. To achieve constant acceleration inside the rocket, the acceleration in the inertial reference frame would be smaller and varying with time as indicated in the graph below. In the graph's title, Acceleration Inside and Acceleration Outside stand for acceleration in the rocket's reference frame and acceleration in the inertial reference frame, respectively.

77

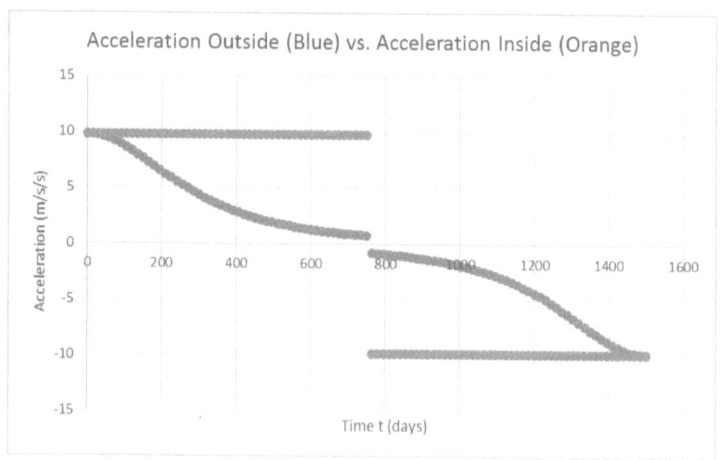

The above graph suggests that accelerating a rocket in the inertial reference frame progressively gets more difficult. In order to reach speed of light, a rocket must accelerate for infinite time or accelerate to infinite magnitude.

Below are the graphs comparing between non-relativistic and relativistic calculations. For clarity when comparing relativistic and non-relativistic calculations in the graphs, the blue represents data points predicted by non-relativistic mechanics and the orange by relativistic calculations.

Beginning with the distance traveled, the relativistic calculations deviate significantly from the non-relativistic ones. Note that the relativistic predictions are accurate descriptions of the real world, while non-relativistic predictions are only good approximations for low speeds as illustrated by the converging points near the origin. The rocket traveled a distance of 2.60 light years within 1500 days in respect to time t, approximately half of the non-relativistic prediction. It is also important to emphasize that while time elapsed more slowly inside the rocket than outside, both non-relativistic and relativistic data points

78

were superimposed onto this graph with respect to time t, not τ. As a result, the curvature of the relativistic set of points is not as strong as the non-relativistic case. As the rocket approaches the speed of light, time dilation becomes stronger and the change in speed becomes less apparent. Thus, there is a linear-like trend in the middle portion of the curve. This linear-like trend also indicates the speed limit c.

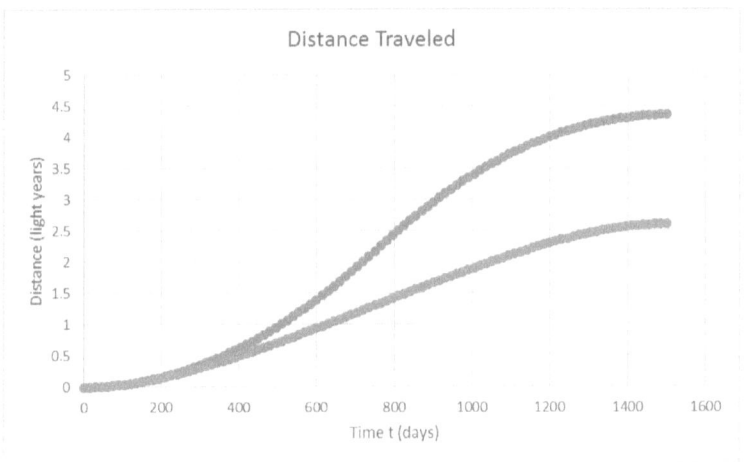

As shown in the next figure, the speeds deviate significantly; this explains the difference of the distance traveled.

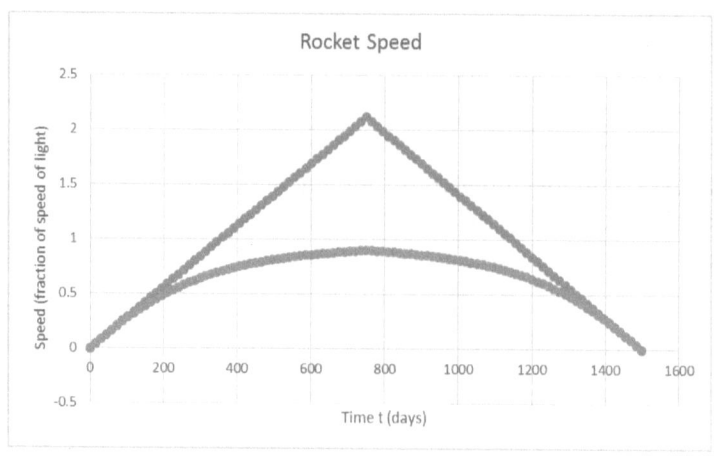

The two graphs above indicate that for a set time period t of 1500 days, the relativistic rocket covered less distance than the non-relativistic rocket. But when new data points were plotted in red so that the same travel distance was covered, time τ elapsed approximately 1306.3 days while time t 1500 days.

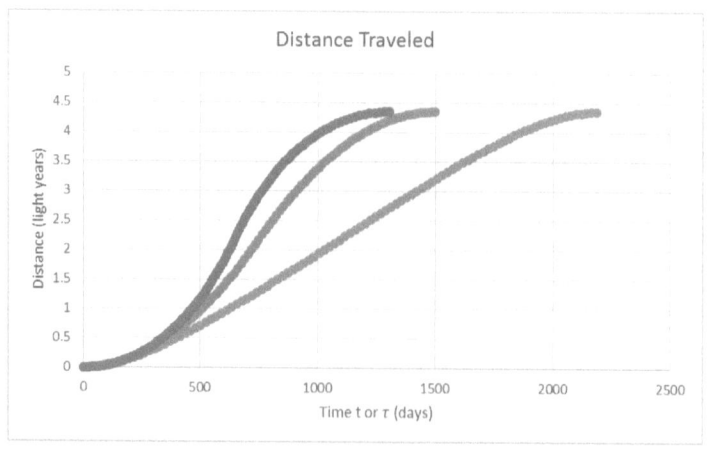

Note that the orange data points indicate that more time was lost for the relativistic rocket in respect to time t. The speed graph below changes as well.

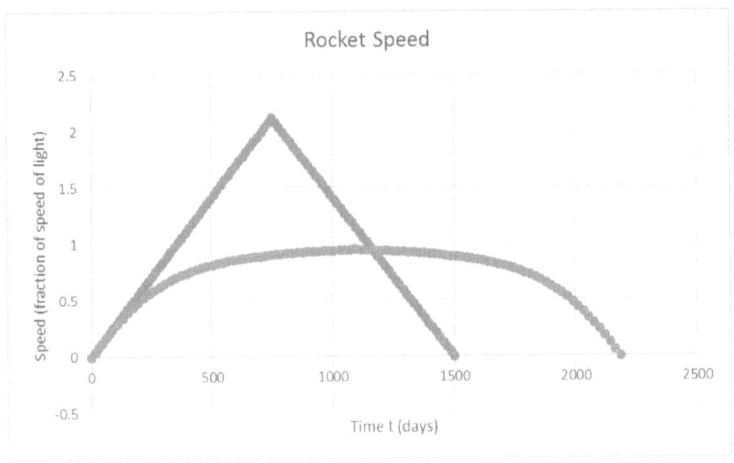

In the graph below plotting mass remaining over time t, the relativistic set of points (orange) do not smoothly decrease exponentially like how the non-relativistic points (blue) do. Instead, there is a linear-like trend for a large portion of the curve, which again correlates to the significant time dilation that occurs when the rocket reaches relativistic speeds.

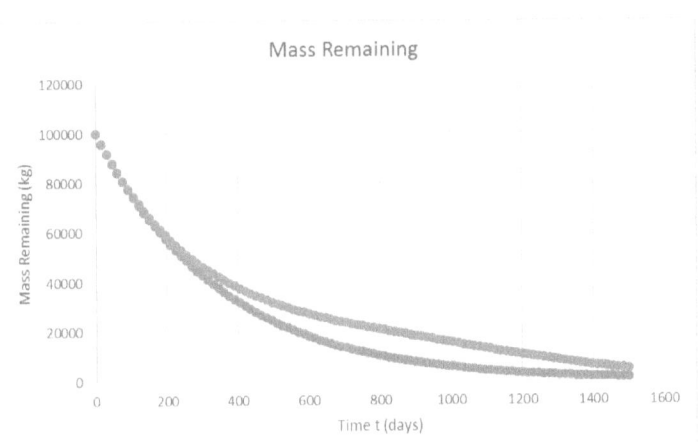

If the mass remaining for the relativistic case is plotted with respect to τ, the following will occur.

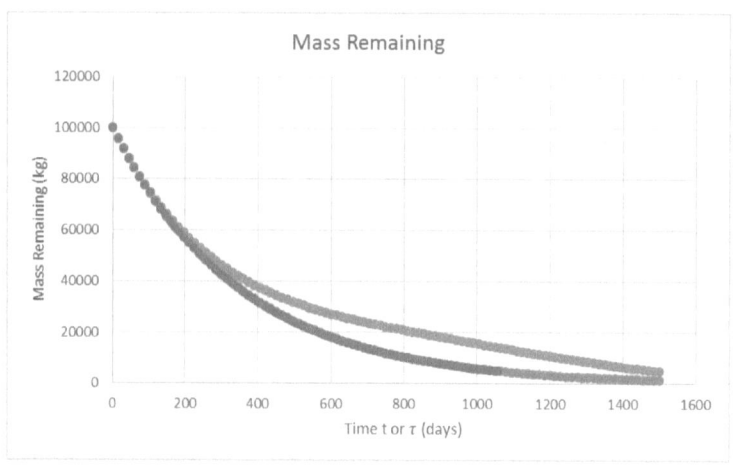

Note that the red displays the rest mass remaining relative to τ and orange and blue relative to t. Because everything is relative, the red should correspond exactly with the blue, which in fact it does as shown in the two curves overlapping.

The mass consumption rate more clearly demonstrates the time dilation effect that also linearized the mass remaining curve of relativistic points. Below is a graph, where orange is relativistic $\frac{dM}{dt}$ plotted against t and blue is non-relativistic $\frac{dM}{dt}$ plotted against t.

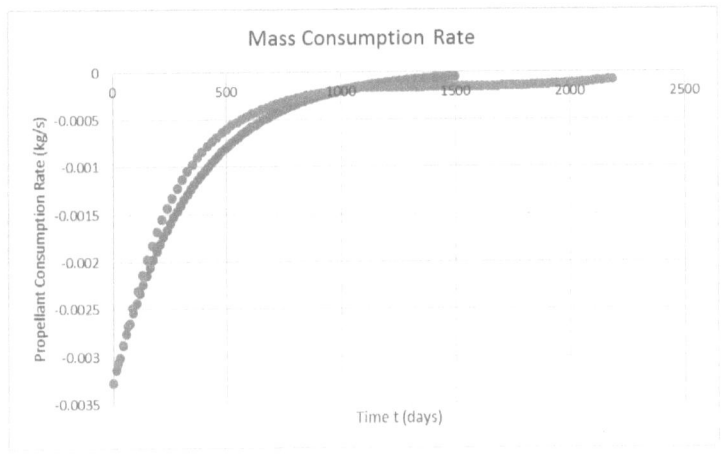

When overlapping the curve $\frac{dM}{d\tau}$ with respect to τ onto the figure above as red, the following graph is produced.

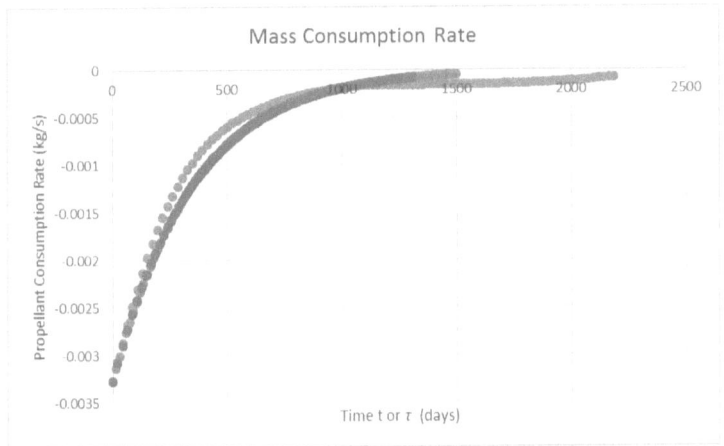

The red data points depicted in the three graphs above indicate that, if there were passengers inside the relativistic rocket, the passengers saved 193.7 days of their lives as well as 1049.4 kg of fuel.

3.5. Conclusion

While observed fundamental properties like mass and time in a rocket approaching light speed behave non-intuitively in an inertial reference frame, they accurately follow the laws set by non-relativistic mechanics inside the rocket's reference frame. This was shown in the graphs displaying rocket mass and mass consumption rate over some time. In addition, when comparing the predictions of relativistic and non-relativistic equations, non-relativistic equations suggest that the rocket is effective in traveling long distances without much expense to fuel. In a real situation however, as indicated by the relativistic equations, it is more difficult for rockets to travel farther for a given fuel due to the upper speed limit c. The one benefit with the relativistic case is the time dilation effect. Because of this phenomenon, when comparing between the time t elapsed for a non-relativistic rocket and the time τ elapsed for a relativistic rocket at a same travel distance, less time and fuel are needed for the passengers onboard the rocket. Thus, though the relativistic effect deteriorates the effectiveness of interstellar travel viewed by observers in an inertial reference frame, it is more beneficial to the passengers in the rocket's reference frame than what the predictions of non-relativistic mechanics indicate.

3.6. Acknowledgement

The author acknowledges Dr. Harold Geller for his constant support and helpful advice throughout the research process. The author also acknowledges the Thomas Jefferson High School Mentorship Program for making this effort possible.

3.7. APPENDIX

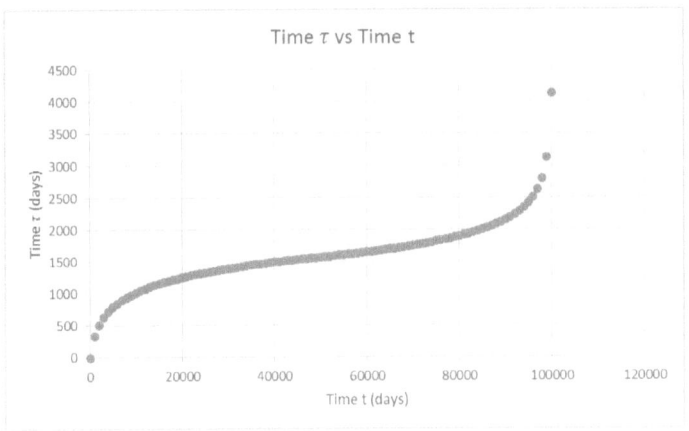

Maximum Speed Reached: 0.999975 c
Total Distance Traveled: 271.9 light years
Relativistic Time τ: 4,126.4 (days)
Projected Time t: 100,000 days
Exhaust Speed: c = 299,792,458 m/s
All mass is propellant: 100,000 kg
Mass Delivered at Destination: 1.25 kg

The two-dimensional graphs shown throughout this chapter can be summarized in the three-dimensional graphs of both accelerating and decelerating relativistic equations relating rest mass (indicated as F for fraction of rest mass), rocket speed (Sr), and relative exhaust speed (Se) below. The input parameter for both acceleration and deceleration plots is 100% rocket efficiency. Another input for the deceleration plots is an initial rocket speed close to the speed of light (2.9*10^8 m/s).

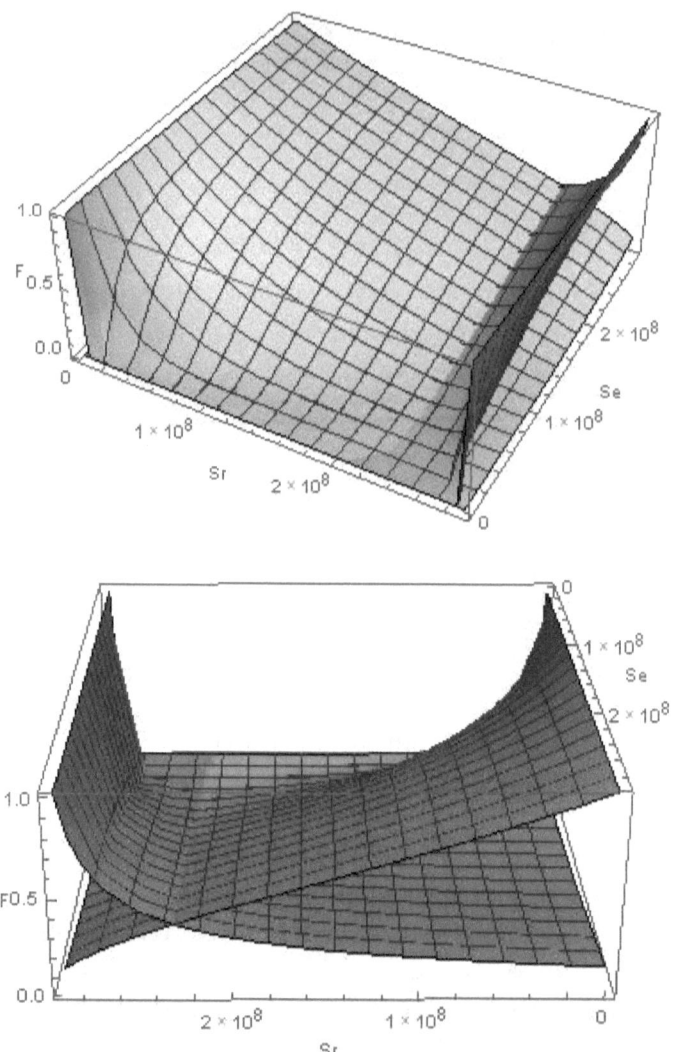

3.8. REFERENCES

Brewer, G. R. (1970). *Ion Propulsion.* New York: Gordon and Breach, Science Publishers, Inc.

Forward, R. L. (1995). A Transparent Derivation of the Relativistic Rocket Equation. *Joint Propulsion Conference and Exhibit* (pp. 1-5). San Diego: AIAA. Retrieved from http://www.relativitycalculator.com/images/rocket_equations/AIAA.pdf

Knorr, B. (2010, June 19). *Uniform Relativistic Acceleration.* Retrieved from Physik: http://www.physik.uni-leipzig.de/~schiller/ed10/Uniform%20relativistic%20acceleration.pdf

Lewis, R. A., Meyer, K., Smith, G. A., & Howe, S. D. (1999). AIMStar: Antimatter Initiated Microfusion For Pre-cursor Interstellar Missions. *AIAA Paper.* Retrieved from http://www.engr.psu.edu/antimatter/Papers/AIMStar_99.pdf

Long, K. F. (2012). *Deep Space Propulsion A Roadmap to Interstellar Flight.* New York: Springer.

Chapter 4: Political Infrastructure
By Alisa Aydin, Peter Lam and Amy Yi

4.1   INTRODUCTION

Establishing a political infrastructure is key to maintaining a social standard and order on the spaceship. There are several factors that go into policy-making. Who is this policy for? Where will this policy be used? What is the end-goal of this policy? Since the multi-generational mission will take hundreds of generations to accomplish, we must do our best to implement a system that will effectively run the spaceship. Rules and regulations are a necessity to any community. Without a political structure, there will be no form of leadership, and an unorganized leadership will jeopardize the entire mission.

In this section, the spaceship's political, judicial, and economic systems will be discussed. First, what type of political system will be utilized on the ship? We concluded that the ideal solution is a combination of democracy and oligarchy. Second, how can we organize the government in a way to prevent corruption and tribalism? The government will create social contracts that explicitly state working conditions for the crew. Third, what kind of criminal justice system should we establish? We decided that the ship should emulate the United States criminal justice system with appropriate modifications to accommodate the unique environment. Fourth, how will we enforce the laws? We concluded that there should be a single, collective, armed police force on board. Fifth, should we use a cash currency? We concluded that it would be difficult to have the same economic system as societies on Earth since we are operating on limited resources. We concluded that the spaceship would not survive long with a monetary economic system. Lastly, should there be an award system in place? Rewards will help motivate the crew while averting dangers of social classes.

4.2   CURRENCY AND AWARD SYSTEMS

Throughout the history of human life, some form of currency and economic system existed in societies. From the hunting-and-gathering societies and bartering systems to the cash and credit, exchange was always existent. An economic system is a factor for organization and productivity in a society, so it is important to consider what type of currency should be used in a limited environment such as the spaceship.

For this mission, it was decided that a cash currency would hinder the spaceship's environment. Therefore, we recommend the implementation of an award system, which will be discussed later, rather than a cash currency.

A cash currency has the benefit of positively affecting an individual's behavior. According to social psychological studies, the neurotransmitter known as dopamine is shown to be active during monetary reward simulations (Izuma et al. 2008). Dopamine is connected to the working memory training and functioning process (Izuma et al. 2008), so this would allow inhabitants to work these crucial areas in their brains. A part of the human brain known as the insula becomes active when faced with risk and loss aversion, which are common situations an individual would face when dealing with currency. The insula signals feelings such as cravings, admiration, love, and indignation (Reyna and Huettel 2014). In an enclosed environment without any interaction with the outside world, it is important for inhabitants to stimulate their brains or it will atrophy.

Despite these psychological benefits, a cash currency is not ideal for a community that is living on finite resources and limited space. Immediate gratification is preferred by biological and psychological factors (Beran and Evans 2012). Self-control holds a crucial role in human behaviors, but there is a greater temptation towards the idea of immediacy and impulsivity over overall value and patience. This can be seen in current society with gambling, shopping sprees, and poor financial planning. A cash currency would allow this system to propagate within the spaceship's economy. It would be best to avoid this situation due to limited resources. Although cash currencies allow for the establishment of income, it is not necessary or sensible to pay individuals on the spaceship because there is no opportunity to increase their resource supply. While their finite resources cover basic needs such as food and water, there is not enough to spare on a system that allows individuals to purchase resources with their earned income. As discussed in Long Term Economic Significance of Currency and Trade Restrictions, more money from income leads to more spending on non-essential goods (Lenschow 1950). This leads to inefficiency in a closed economy.

Another reason to stay away from a cash currency system is due to motivation hindrance. It was found in a study that extrinsic motivation, known as monetary rewards in this scenario, may conflict with the individual's personal, intrinsic motivation (Benabou and Tirole 2003). Once a person realizes that they are able to receive a monetary payment for a task, they will no longer be motivated to complete that same task for free. If the spaceship's inhabitants acted this way, the mission would surely fail. Inhabitants will have their own jobs on the ship, and every single person will be an invaluable asset. If one decides to stop putting effort into their work, the whole mission can be jeopardized. In order for this to be prevented, inhabitants should not be given monetary rewards.

For these reasons, and others, we concluded that a cash currency should not be used for the duration of this mission. Despite the numerous factors to consider, a monetary system has a higher probability to create class tensions between inhabitants. It is difficult to determine how resources will be allocated with a cash currency system. Instead of worrying about the possible establishment of class systems and monopolies over resources, we will implement an award system for our economy.

More than hundreds of people will inhabit the spaceship, so an organized system must be implemented to keep everything in order. Both social and monetary rewards affect an individual's behavior, so it is imperative to consider how to cater policies to this unique environment. Since the population on the spaceship is all there is, the unity of the group is crucial to the success of the mission. An award system that uses points will help keep inhabitants motivated while decreasing chances of conflict.

Cash currency and award systems are relatively the same when it comes to the effects on human's brain and behavior. Similar to how monetary rewards active dopamine; rewards stimulate the brain as well (Izuma et al. 2008). For this mission, an award system will be implemented to maintain a steady level of motivation, psychological stability, and competition. It is important for the spaceship to have a system that promotes improvement and innovation because of their limited surroundings.

Reputation significantly influences cooperation rates (Izuma et al. 2008). In an economic system where there are wealth inequalities, there would be an immense amount of judgment within the group. To keep negative reputation based on wealth out of the spaceship, the award system should not be treated as a monetary currency. If certain social gaps are eliminated on the spaceship, then each person's reputation will be more leveled. When one is viewed in a positive light by others, they tend to cooperate more than if they are viewed negatively. An overall conflict-free environment should lead to higher cooperation rates between groups. Cooperation is beneficial to the mission since everyone must work together in order to successfully live in a confined area while striving to reach a common goal.

An individual that is risk averse in one domain is likely to be risk averse in another (Reyna and Huettel 2014). Domains include categories such as financial, social, ethical, and recreational. Having an award system can initially expose inhabitants to risk aversion so that they are less likely to make risky decisions that may damage the ship's mission.

Rewards are not necessarily monetary rewards. Awards are claimed through points which are given based on performance in both the working and living areas. Performance in the workplace is gauged by productivity, and can be confirmed by the department's leader. Responsible and ethical behavior on the ship will result in awarded points as well. On the other hand, wrongful and violent acts can lead to point deductions based on the sentence passed down from our criminal justice system. The deduction of points will be taken by percentage and not by integers to ensure fairness. Redeemable awards may include items or entertainment. The purpose of these awards is to provide incentives for inhabitants for the duration of the mission. However, all awards must take the spaceship's environment and limited resources into consideration.

Points from this award system can be called currency since it will be used as a measure of value. Currency is defined as a process of exchange through any medium (Temple 1899). Since points can be traded in to redeem different awards, it acts as a substitute for cash currency. Despite the similarities between cash and points, the small differences will make a difference in the making of the spaceship's economic system and solidarity.

4.3   LAW ENFORCEMENT AND JUSTICE SYSTEM

It has been historically shown that community-formations are dependent on law and law-enforcing authority with established machinery (Reith 1945). It is thus vital for our ship to implement a criminal justice system that meets the expectations for criminal law and its fair application along with a well-equipped, authoritative force to secure the observance of laws. These criteria will be met with the following choices: A criminal justice system that emulates the United States' and an armed, collective police force not under the direct jurisdiction of the oligarchy. This paper will explore both choices and our critical need for them over others.

Potential solutions for our ship's criminal justice system consist of giving the Oligarchy judicial review, giving the Oligarchy the ability to instate magistracies with judicial review, or having a court-based system. To begin explanation of why our ship has chosen the third solution, we must first analyze the potential for corruption under each system. Oligarchies with judicial power, such as Athenian oligarchies, have existed in the past with significant abuse of power. Athenian oligarchies have been interpreted as showing more concern towards what was to their advantage than what is just (Bonner 1926). Constitutional powers such as the death penalty and having an overtly self-serving interpretation of the law gave way to the overthrowing of democracy within Athenian Society (Bonner 1926). The same issues arise with the creation of magistracies as they tend to be filled with adherents of the Oligarchy. It would not be ill-founded to assume such corruption could be present on the ship under a similar system.

A court system negates such corruption by disallowing long-term authority to a single collective such as an oligarchy or magistracy. Procedures such as trial-by-jury directly involve members of society by giving them the power to interpret the law. Collaboration of this nature is vital since the efficiency of a justice system is gauged through societal response and satisfaction (Nettler 1979). Nettler mentions that there is "a universal (i.e. cross-cultural) consensus concerning the disapproval of a number of crimes." With that said, our culturally diverse crew would function at a level adequate enough to apply criminal law deemed just by the community

This separated branch also allows for a checks and balance system between it and oligarchy (Corbett 2012). It is recommended that judges be replaced by oligarchs that are on a revolving schedule, so any assessment of questionable Oligarchy actions would be done through a jury trial instead. A significant, uneven number of randomly selected crew members will evaluate the oligarchy's use of extralegal powers. The inconveniences of an ill-informed public would be negated by the trial which will reveal all relevant information. This method was deemed best since extra legalism relies upon popular judgment of executive discretion (Corbett 2012).

With changes such as the replacement of judges, it is vital to be aware that our ship's implementation of a criminal justice system only emulates the United States' to a certain degree. This is due to strict limitations on population and resources. Harsher penalties, particularly the death penalty, will be given in place of long-term sentences. High offences or repeated smaller offences will result in termination. It has also been seen that criminal law distinguishes among classes of citizens such as children and females (Nettler 1979). No such distinctions will be made within the ship. Trials will not reveal age, gender, race or any other definable specific so that all penalties are applicable to all persons. Alterations such as this seek to optimize the ship's resources while ensuring uniformed application of criminal law. These mentioned and unmentioned changes exist for the purpose of contributing to not only a functioning society, but a resource-aware one as well.

It should be assumed that law enforcement will exist on the ship. Law enforcement in this case refers to a police force rather than a military force, primarily because military forces have never been successful as an instrument for enforcing laws. "Limitations of the structural organization of armies, the need of keeping military formations in concentrated form in order to maintain their power, and the impossibility of applying military principles to the conduct of internal law-enforcement," make military force unsuitable for the ship (Reith 1945). Any option against the use of police and for the use of moral force alone to enforce law is unrealized. Moral force has rarely, if ever, been known to be independent of other forces (Reith 1945). The very idea of police is a pre-eminent factor in the preservation of law and order in democratic and similar societies (Junior and Muniz 2006). With that said, the ship must have a police force, but to what degree should be discussed.

The primary consideration for law enforcement is largely dependent on organization and the capacity to use force (Junior and Muniz 2006). Two potential methods for organization are to have a single, collective police force or to have multiple independent police forces. Due to the restricted population of less than 500, the division of an overall unit of 10 or so police into multiple departments is nonsensical. Staff would be lacking for both investigation and response while tribalism could easily occur between the departments (Colby 1978). One department is therefore more operative.

Significant issues that afflict single, large departments are negated by the population restriction. A major issue large departments face is the lack of cooperation between them and the public due to unfamiliarity with the area and locals (Smith 1933). Such cooperation is essential to the success of the police. Fortunately, a small overall population and ship-restricted space will resolve this issue. It is stated by Smith that out of the thousands of police agencies in the United States, the number of able police administrators does not exceed ten or so men. With that in mind, the ship's police department will likely have both small and large department advantages such as access to needed facilities, knowledge of the area, and a good relationship with the public (Colby 1978).

Police have a unique ability to compel in situations in which force may be needed. This "use of force" by police encompasses not only the physical aspect of force but the potential use of it as well (Junior and Muniz 2006). It is then vital to give police the proper tools to ensure they have the predisposition as well as the capacity to use force. An appropriate loadout for the police force on the ship should include conventional firearms with frangible ammunition as well as less than lethal weapons such as Tasers. Other firearms such as snipers will be limited to circumstantial use. Weapons will also be available for non-police use if mutiny occurs among the whole police force.

It is imperative that justice be deemed rational and satisfactory by the public. Rational justice is non-objective in that "one will always come up against an irreducible vision of the world expressing nonrational values and aspirations" (Nettler 1979). Thus whatever justice apparent on the ship must reflect public opinion in order for the system to be gauged as effective. The suggested methods for law enforcement such as the court system should not be assumed to be the most proficient method for the ship at all points of the mission, especially within a multigenerational one. Societal values will undoubtedly change and so must the justice system in order to remain pragmatic.

4.4   GOVERNMENTAL SYSTEMS AND CORRUPTION CONCERNS

Creating a political infrastructure is one of the key components to establishing a social and political expectation, or "social norm," of sorts. Without a clear government, the citizens among the ship will not have any organization. A state of anarchy is a recipe for disaster. The ship cannot expect to be successful and survive multiple generations if it cannot regulate its people on board. A political infrastructure needs to clearly promote values and expectations that will create and maintain a civil environment.

There are numerous options as to which type of government could be selected. Historically speaking, there have been cases of anarchy, monarchy, democracy, fascism, oligarchy, and theocracy, just to name a few. For the purposes of this mission, we have selected to combine democracy and oligarchy into one hybrid-like government. Oligarchy has been incorrectly misconstrued as incompatible with democracy (Page and Winters, 2009). We believe that the two forms of government can be used together to set the ship up for success.

Anarchy has already been ruled out. It is not sensible to run a mission without any type of government. There would be no way to efficiently establish any rules on the ship and there would likely be a lot of crime and corruption. This would be a sure-fire way to have chaos on the ship.

Governments such as a monarchy, fascism, or theocracy have all been documented well in history. Although these forms of government have the positives of a strong individual leader, this may pose threats to society if the leader becomes unpopular and does not interact well with the population. History may repeat itself and the society may overthrow or branch away from the leader.

Also, it is preferred that the ship should have religious freedom, so a theocracy should be avoided in order to keep political and religious power separate.

Oligarchies have been successful on occasions in the past however are not that frequently demonstrated. Oligarchies can successfully represent its people much like a democracy, however there is a danger of oligarchs becoming corrupt. Oligarchs can sometimes become corrupt usually due to economic reasons.

In the modern world, it is safe to say that democracy is one of the most successful and efficient forms of government. It allows the voice of the people to be heard and represented. Majority rule is also used which is a universally recognized and accepted ideal. We are striving to successfully combine these democratic ideals within an oligarchy. An oligarchy will fit well with this mission because on the ship there will be many different types of departments and occupations. The leaders of these groups will serve as oligarchs, yet rule under the democratic principles.

When the initial first generation of the trip is established, each individual department will select a leader. There will be a leader from the engineers, teachers, chefs, military, etc. Each voted leader would then serve as the oligarch for their specific concentration and be a representative of its people. The key to selecting oligarchs is that the oligarchs must appropriately match the population. If there are too many selected then power is not effectively distributed (Payne, 1968). Oligarchs would serve a term of 6 years; they may be reelected one time. This will help prevent against corruption and maintain that they are always trying to better the society on the ship and promote the values of their own department. There needs to be a clear, strong, leadership role for the oligarchs because if everyone on the ship is equal then the society will not effectively develop; there needs to be certain individuals with power (Raths, 1954). The oligarchs will meet once every two weeks to get a status update on the ship. Oligarchs are responsible for organizing meetings within their own population to be up to date with what the majority of people are thinking on certain issues. When a law or decision needs to be made, the oligarchs will have a vote amongst themselves and a majority rule basis will be utilized.

When dealing with a government that involves the principles of an oligarchy and a democracy, it is necessary to eliminate corruption and tribalism, and crime altogether, as much as possible because any of these activities are detrimental towards the ultimate goal of the ship and serves as a distraction. The government needs to operate as efficiently and fairly as possible in order to guarantee a positive environment in the ship. If everybody on the ship is content with how things are going then they will all be motivated to do their task to the best of their ability. If corruption is present, certain departments on the ship will be receiving unfair treatment and benefits, which will throw off the balance of the whole ship.

Interestingly enough, sources say that some crime is actually beneficial and may enhance the cohesiveness of the group (Scott, 1976). Crime brings together upright consciences and concentrates them. It is almost a backwards way of establishing a norm: it tells society what not to do. That being said, the ship still needs to do its best to limit crime and corruption as much as possible.

One key thing to keep in mind is the population size that an oligarch will serve for. Oligarchs have more power in a smaller group, and less power in a larger group (Krauth, 2006). Oligarchs need to be selected in departments that are not too small so that the oligarch can abuse power, but not too big so that they are not powerful enough.

Also as a whole, the ship needs to establish a set of

norms that everybody will live by. There first needs to be an integrative social contract theory, which basically implies that the government will respect those that it governs. Next there needs to be a macro social contract; this contract is a set of rules that, for our purposes, the oligarchs will agree upon. Finally there needs to be a micro social contract; it helps prevent corruption and establishes a minimum level of conduct (Nichols, 2009).

Although some sources say that crime is necessary to establish certain norms, it would be best for the ship to establish these norms in a different way. In such a concentrated mission, corruption and crime need to be avoided altogether as much as possible to keep all tasks focused.

Oligarchs indeed need to be in charge of an appropriate amount of people. If there are departments that only have a very small amount of people (comparatively), then multiple small departments can be combined in order to have a sensible size. Vice versa, if there is a department that is very large in size, perhaps more than one oligarch can be appointed for that specific field.

The integrative social contract theory and macro and micro social contracts are a good idea and will help establish these social norms among the government and the society. The one potential problem is the oligarchs may corrupt together and create a contract that may benefit all the oligarchs and not help out the general population.

The government is required to create all three social contracts: the integrative social contract, macro social contract, and the micro social contract. In order to make sure that the oligarchs are still not being corrupt, once the contracts are made they will be presented to the public. If there is a majority of disapproval from the society, then the contracts will be forced into revision.

In addition to the three social contracts, there are several factors that contribute to an effective government that is not corrupt. One of the key issues is currency. This ship will not have currency. Historically speaking, oligarchies usually fail due to oligarchs being bribed and becoming too powerful (Fishkin and Forbath, 2014). A civilization without currency, based on a merit based award system, will have an "even playing field' of sorts and require that all people on the ship work together equally and fairly to all reach the ultimate goal of the ship.

Social and moral integration are also major factors in forming strong relationships. The oligarchs need to stay connected with their department as frequently as possible. There will be a ship-wide rule that oligarchs meet with their people at least once every two weeks. If they wish to meet more then they may. If each department can maintain a strong and positive level of functional, normative, and interpersonal integration, then all the departments together (and their oligarchs) will work cohesively and effectively.

4.5 REFERENCES

Benabou, R., & Tirole, J. (2003). Instrinsic and Extrinsic Motivation. The Review of Economic Studies, 70(3). Retrieved October 6, 2014, from JSTOR

Beran, M., & Evans, T. (2012). Language-Trained Chimpanzees (Pan troglodytes) Delay Gratification by Choosing Token Exchange Over Immediate Reward Consumption. American Journal of Primatology, 74 (9), 864-870.

Blumstein, A., and R. Larson. "Models of a Total Criminal Justice System." Operations Research 17.2 (1969): 199-232. JSTOR. Web. 1 Dec. 2014.

Bonner, Robert J. "Administration of Justice under Athenian Oligarchies." Classical Philology21.3 (1936): 209-17. JSTOR. Web. 27 Nov. 2014.

Brosnan, S., & Waal, F. (2004). A Concept of Value during Experimental Exchange in Brown Capuchin Monkeys, Cebus apella. Folia Primatologica, 75(5), 317-330. Retrieved October 6, 2014, from Proquest.

Corbett, Ross J. "Suspension of Law during Crisis." Political Science Quarterly 127.4 (2012): 627-57. JSTOR. Web. 1 Dec. 2014.

Dib, A. (2011). Monetary Policy in Estimated Models of Small Open and Closed Economies. Open Economies Review, 22(5), 769-796. Retrieved October 6, 2014.

Fishkin, Joseph and Forbath, William. The Anti-Oligarchy Constitution. 94 B.U. L. Rev. 669 (2014)

Hira, A. (2010). The evolutionary patterns of political economy: Examples from Latin American history. Journal of Bioeconomics, 12(1), 1-28. Retrieved October 6, 2014.

Houston, A., Fawcett, T., Mallpress, D., & McNamara, J. (2014). Clarifying the Relationship between Prospect Theory and Risk-sensitive Foraging

Theory. Evolution and Human Behavior. Retrieved October 6, 2014.

Izuma, K., Saito, D., & Sadato, N. (2008). Processing Of Social And Monetary Rewards In The Human Striatum. Neuron, 58(2), 284-294. Retrieved October 6, 2014.

Junior, D. P., and J. Muniz. "'Stop or I'll Call The Police!': The Idea of Police, or the Effects of Police Encounters Over Time." British Journal of Criminology 46.2 (2006): 234-57. JSTOR. Web. 1 Dec. 2014.

Krauth, Brian. The Canadian Journal of Economics. Vol 39, No. 2 (May, 2006), pp. 414-433

Letiche, J. (2006). Positive Economic Incentives: New Behavioral Economics and Successful Economic Transitions. Journal of Asian Economics, 17, 775-796. Retrieved October 6, 2014.

Lenschow, Gerhard. "Long Term Economic Significance of Currency and Trade Restrictions." The Canadian Journal of Economics and Political Science 16.1 (1950): 63-69. JSTOR. Web. 6 Oct. 2014.

Nettler, G. "Criminal Justice." Annual Review of Sociology 5 (1979): 27-52. JSTOR. Web. 1 Dec. 2014.

Nichols, Philip. Journal of Business Ethics, Vol. 88, Supplement 4: A Tribute to Thomas W. Dunfee a Leader in the Field of Business Ethics (2009), pp. 805-813

Page, Benjamin and Winters, Jeffrey. Perspectives on Politics, Vol. , No. 4 (Dec, 2009), pp. 731-751

Payne, James. World Politics, Vol. 20, No. 3 (Apr., 1968), pp. 439-453

Peter, Colby. "Small Police Forces in a Large Metropolitan Area: Can They Do the Job?" State & Local Government Review 10.1 (1978): 28-34. JSTOR. Web. 30 Nov. 2014.

Raths, Louis. Journal of Educational Sociology, Vol. 28,

No. 3 (Nov., 1954)

Reith, Charles. "Comparative Systems of Law-Enforcement." Transactions of the Grotius Society 31 (1945): 150-73. JSTOR. Web. 1 Dec. 2014.

Reyna, V., & Huettel, S. (2014). Reward, Representation, and Impulsivity. The Neuroscience of Risky Decision Making.

Scott, Robert. Social Forces, Vol. 54, No. 3 (Mar., 1976), pp. 604-620

Smith, Bruce. "Politics and Law Enforcement." Annals of the American Academy of Political and Social Science 169 (1933): 67-74. JSTOR. Web.

Temple, R. (1899). The Beginnings of Currency. The Journal of the Anthropological Institute of Great Britain and Ireland, 29(1/2), 99-122. Retrieved October 6, 2014, from JSTOR.

Chapter 5: Starship Protection

By Heather Gonyeau, Rachel Rockrohr and Katelyn Squicciarini

5.1   INTRODUCTION

A spaceship travelling many light-years poses many dangers to the humans within it. These hazards include but are not limited to cosmic debris, disease, radiation or even threats from other humans.   This section will discuss possible technologies and systems to protect the spaceship and those inside of it.

The first section focuses on defensive shielding mechanisms. With a ship traveling at such high speeds, even small particles of interstellar dust would be capable of piercing the ship's hull. Additionally, space contains harmful high-energy particles that would lead to adverse health effects for the crew. Instead of using two different technologies to solve these issues, a magnetic/electrostatic plasma shield that would offer both defense from radiation and debris would be the most advantageous solution.

The second section discusses two subjects.   The first half of the section focuses on the different offensive weapons that can be used to protect the spacecraft against space debris. The ideal weapon will take up little space, weight, and energy while still sufficiently protecting the spaceship in case of an emergency. The second half of the section focuses on which handheld weapons should be used inside of the spaceship. Handheld weapons are meant to be a precaution against possible assailants aboard the spaceship.   Weapons will be limited to non-lethal weapons that should only be used as a last resort.

The final section focuses on protection from germs, disease, and other passengers. Pathogens grow and react much quicker in space and are more harmful. It is important to know how certain pathogens will react and how people's careers can have an effect on the spread of disease if it becomes an issue. When other passengers become harmful to not only the mission but also other passengers it is important to have a plan for how to deal with those situations. Contrary to other ideas, the section focuses on the most ethical way of executing people in order to prevent them from suffering. The process would start with quarantine from the rest of the population and then move to administering anesthesia and finally expelling individuals into space while unconscious in order to prevent suffering.

5.2 DEFENSIVE SHIELDING SYSTEMS

In addition to protection on the inside of the spaceship, thought must be put into how to protect the ship and its inhabitants from the hazards of interstellar space. Two of the most significant issues with interstellar travel are radiation poisoning and cosmic debris. Although it was originally assumed that different methods of protection would be needed for each cosmic danger, it is in fact possible that one solution would overcome both these obstacles. After investigating multiple modes of protection, my proposed solution would be to surround the manned areas of the space ship with a magnetic/electrostatic plasma shield.

The cosmos are a perilous environment for humans. Interstellar space is permeated with high-energy particles. These particles are part of the Galactic Cosmic Ray (GCR) background radiation that includes nuclei of elements from hydrogen to iron.

These particles, along with other high energy protons from solar flares (SPE), are extremely hazardous. Earth's magnetic field protects organisms from damage by radiation. When a spaceship moves beyond this defensive bubble, crews are exposed directly to GCR and SPE radiation (Schimmerling 2011). Ship inhabitants would overtime develop a variety of illnesses including Acute Radiation Syndrome. Prolonged, intense exposure to radiation would result in damage to the hematopoietic, cardiovascular, and central nervous system (Seed 2011). After multiple generations of radiation exposure, offspring are likely to develop serious genetic defects as witnessed in the offspring of victims of the Chernobyl disaster (Weinberg 2001). Therefore, it is of extreme importance to shield the crew from cosmic radiation.

An initial proposed solution to combat this issue was creating an ice shield that would surround the ship. An experiment done by scientists in 2010 tested a "curtain" of dispersed water droplets as a radiation shield (Collin et. al. 2010). Although this idea proved to be a viable option for a colony, say on a Mars-like planet with no protective atmosphere, a water curtain would likely freeze in the harsh temperatures of interstellar space. There is not enough research done on how to create or sustain a ship based ice shield for it to be considered a plausible solution for this mission.

Another potential hazard of interstellar travel is cosmic debris. In addition to dangers faced by asteroids and larger debris, at the intense speeds required for interstellar travel, even small particles would impact the ship with enough force to tear through even a steel hub. This will not only be a danger to the crew and equipment but it would take an enormous amount of personnel to be constantly repairing the damages. In order to protect ourselves from this hazard, it is necessary to have a shield that can block small debris particles (larger debris such as asteroids are covered in the spaceship weaponry section). In this case, a magnetic shield would be useful because it would dispel charged particles away from the ships hub.

Scientists are in the process of testing a magnetic shield that would defect the energetic ions that are hazardous to manned space flights. Their research is to determine the effectiveness of a magnetized plasma barrier in expelling an impacting, low beta, supersonic flowing energetic plasma representing the solar wind. The proposed "mini-magnetosphere" works via a dipole-like magnetic field and a plasma source. Tested in the lab via particle-in-cell hybrid simulations using kinetic ions and fluid electrons, the results reveal successful simulations on a small scale. With the success of these primary experiments, researchers can move forward with testing how a magnetosphere with an expanded magnetic sphere will react in a space plasma environment with the presence of a solar wind. The following quote explains the proposed design of this technology, "in these experiments, a plasma beam, guided by an axial magnetic field in a cylindrical linear chamber, hits a dipole magnetic field created by a permanent magnet. Preliminary experimental results already reveal the formation of a very sharp shock structure, thus indicating the formation of a mini-magnetosphere" (Gargaté et. al. 2008). Much like Earth's own magnetic shield protects the planet from harmful radiation, this magnetosphere would surround the spaceship and guard it from interstellar rays. The following diagram illustrates these interactions.

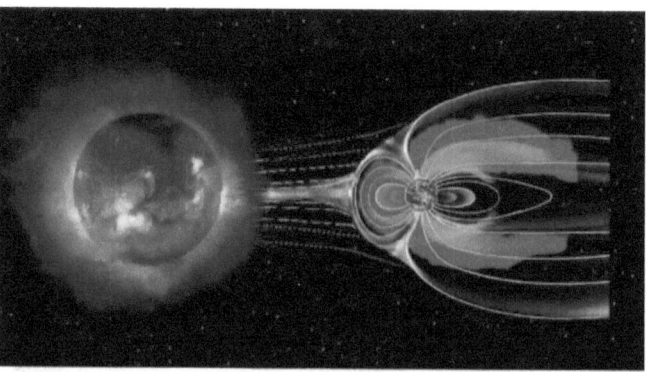

Source: NASA (http://sec.gsfc.nasa.gov/plasma_fountain)

The hot plasma would also vaporize any neutrally charged particles that were not dispelled by the magnetic shield, thus solving the issue of cosmic debris. The shield can be fueled by the same sustainable energy source that is used to power the rest of the ships operations, thus it does not require many additional resources. In another experiment, researchers tested how flowing energetic plasma representing the solar wind interact with a dipole magnetic field. The results demonstrated the potential for creating a small hole in the plasma where a ship could, in theory, safely reside. "Together the experimental and modeling results demonstrate the pivotal role of particle kinetics in determining the plasma transport barriers expected in artificial mini-magnetospheres." (Bamford et. al., 2008) Although more research needs to be done to extend these results to the proper scale, it shows that this structure is a feasible option for protection.

In conclusion, a magnetic/electrostatic plasma shield is the best defense against the hazardous conditions of interstellar travel. More research needs to be done on expanding the shield to the right proportions to cover and protect a large spaceship but since this one technology will solve both the issue of interstellar debris and cosmic rays, it will save considerable amounts of time and money over if two separate defenses needed to be developed. With a magnetic/electrostatic plasma shield, our spaceship could safely arrive to Kepler 186f with no damage to the hull or crew.

5.3   SPACESHIP WEAPONRY

Offensive weaponry is a necessary precaution to have aboard the spaceship because there may be unforeseen situations such as asteroids heading towards the spaceship without time to maneuver out of the way. Added defense will also give the people on board the spaceship the peace of mind of knowing that the offensive weaponry can be used in emergency situations. If at all possible, offensive weaponry should be used as a last resort in order to preserve resources and minimalize possible collateral damage. Along with offensive weaponry attached to the spaceship, offensive weaponry is also necessary inside of the spaceship. While ideally everyone on board the spaceship will stay nonviolent and behave civilly, trusting that this situation will be reality is a risk that cannot be taken. Certain people could become unstable, or groups could become riotous, and security personnel aboard the spaceship need to have weapons in order to protect the general public and maintain the integrity of the spaceship.

There are several options when it comes to offensive weaponry on spaceships. While none have been tested in interstellar space, some may be taken on board the spaceship and developed there. Deblois et al. 2005 identify three kinds of spaceship weaponry: Directed energy weapons, kinetic energy weapons, warheads delivered from space via a Common Aero Vehicle, or microsatellites that act as mines. All of these weapons will be addressed as possibilities for use on the Kepler 186f mission. These weapons will be evaluated to determine which, if any, may be of potential use as offensive weaponry attached to the spaceship.

Electron guns are a type of directed energy weapon. Electron guns shoot out a concentrated electron beam using solar wind as its propellant (Evlanov et al. 2013). A problem with electron guns is that it is hard to maintain the heat necessary to get the thermo-emission cathode to release the right amount of electrons (Evlanov et al. 2013). The heat required would be about 2200K and the heater would last for about 4000 days (Evlanov et al. 2013). This means that a high voltage power supply of between 20 to 40 kV and power up to 2.5kW would be necessary (Evlanov et al. 2013). This may be a problem as such power may not be able to be supplied to the electron guns due to limited space and resources aboard the spaceship. If such power were possible and can be steadily supplied throughout the length of the Kepler 186f mission, than electron guns may be a plausible weapon to take aboard the spaceship and engineer and improve throughout the mission.

A type of kinetic energy weapon is a rail gun. Rail guns shoot out 6 meter long poles made of steel rails charged with 1.5MA currents (Mehlhorn, 2014). They create damage by gouging and melting whatever they come into contact with (Mehlhorn, 2014). Rail guns would require less power than electron beams since they only need the energy to charge the rods with currents and then eject them rather than create a sustained beam of electrons. The problem with using rail guns is that the spaceship would need to carry several hundreds of poles or at least have the ability to make more poles. This would be a drain on resources just like the electron gun would be, as well as added weight onto the spaceship.

The final two weapons, warheads and microsatellites are less feasible than rail guns and electron beams for a number of reasons. Microsatellites are essentially miniature satellites that are meant to destroy other satellites (Deblois et al., 2005). It would take up a lot of space and weight to house several microsatellites, making them not ideal for the mission. The fourth type of weapon, warheads delivered in space, is not a good option for the Kepler 186f missions because they require more resources than rail guns and electron guns as not only would the spaceship need to be able to house multiple warheads it would also need to be able to house small shuttles that can deliver the warheads (Deblois et al., 2005). Ultimately, electron guns or rail guns seems to be the most feasible weapon to use aboard the spaceship. Choosing between using electron guns or using rail guns may come down to a matter of which one is less draining on resources and space. Electron guns require power and heat, while rails guns require power and physical rods to use as projectiles. Resource management will have to determine which weapon is the least taxing on resources.

There are several types of offensive weaponry that can be used aboard the spaceship. Projectile weapons such as handheld guns will not be used aboard the spaceship in order to avoid damaged to the ship and possible bullets ricocheting off of walls and hitting unintended targets or machinery. Instead, non-lethal weapons, also known as less-lethal weapons, will be proposed as an appropriate physical method for maintaining safety. Handheld weapons are only meant to be used as a final resort, and only when the user deems that there are no non-violent methods that can be used to diffuse a situation. The United States Department of Defense defines non-lethal weapons as weapons meant to incapacitate with the least physical damage possible (Rappert & Wright 2000). Non-lethal weapons include kinetic energy weapons such as batons, conducted energy devices (CEDs), and chemical weapons such tranquilizers (Downs 2007). The use of non-lethal weapons aboard the spaceship should fall into three tiers: batons, CEDs, and anesthetics, respectively.

Batons should be the first choice for physical control after all other non-violent methods fail. Batons allow more control as they are short range weapons that can be effectively used to control one assailant. Security personnel should use batons to incapacitate an assailant with as little physical damage as possible. Non-lethal weapons may be used ethically only if they give the security personnel more time and space to make good decisions in violent situations, reduce the suffering of non-assailants, help lead to peace, and lower the amount of casualties (Kaurin 2010). If a baton is used to continually beat someone who is already submissive to security personnel than that baton is no longer a non-lethal weapon. Security personnel need to be properly trained to use all weapons with caution and hesitation.

When batons fail, or are deemed impossible to diffuse a situation, then CEDs should be the next weapon security personnel use. A common type of CED, Tasers, use thin wires attached to small darts to latch onto a person and incapacitate them with an electric shock (Schneider 2005). Tasers should be used with extra caution as they may be accidentally misfired at a bystander when they were meant for an assailant. Once again security personnel need to exercise caution and safety when using Tasers. A problem with Tasers is that they may be misfired onto a piece of equipment that may be damaged by the electric currents. Depending on where in the spaceship the Taser may be used also determines whether or not they are viable for use.

A final resort for security personnel should be chemical weapons. A common chemical weapon is a canned gas irritant such as pepper spray (Downs 2007). The problem with using chemical weapons such as pepper spray is they will be released into a small confined space. While security personnel may wear gas masks, bystanders would be unavoidably affected by the irritant. Also, the irritant could get into the vents and spread throughout the spaceship. While measures can be taken to avoid this situation, for the safety of the spaceship and its inhabitants, gaseous irritants should not be used aboard the spaceship. Instead, anesthetics, such as tranquilizers, should be used. Anesthetics avoid the problem of air contamination. Furthermore, if they are misfired onto an object in the spaceship they will not do the same damage that a bullet will. They can be used by hand or as a projectile, offering flexibility. They should only be used as a last resort as chemical resources aboard the spaceship will be limited, and anesthetics for the use of the violence control is less necessary than anesthetics for the use of pain control during medical procedures.

Ultimately, the safety of both assailants and bystanders lies in the hand of security personnel. A major problem with non-lethal weapons is that they are subject to the judgment and handling of their user just as a lethal weapon is (Kaurin 2010). Just as with any job aboard the spaceship, security personnel will need to be trained to understand when certain weapons should be used, and to be able to make good judgments in unclear situations. Ideally, non-lethal weapons will never need to be used aboard the spaceship, and all disputes will be solved civilly. Nonetheless, non-lethal weapons are a necessary precaution meant to protect the safety of all people aboard the spaceship. Weaponry attached to the spaceship is also a necessary precaution, although ultimately all types of weapons on the spaceship should only be used as last resorts in emergency situations.

5.4   PROTECTION FOR HEALTH IN SPACE

When traveling in space one of the most important things to look at is the protection of the travelers. One of the most important aspects of protection is health. Knowing how germs react, what space does to a human's body, as well as protecting patrons on the spacecraft from individuals from a multitude of issues that may arise is vital to making the mission successful. Patrons will need protection not only from germs and those that have contracted a disease but also from other patrons that prove to be a threat to others as well as the overall mission due to violent tendencies.

One study was done that looked at disease and germs in simulated microgravity (SMG) and the effect of microgravity (similar to gravitational effects in space) on microorganisms. The study showed that microgravity made the microorganisms act more efficiently as pathogens. "Salmonella enterica serovar Typhimurium cells, grown under SMG conditions for a mere 5-10 hours, kill mice more rapidly than do such cells grown under ordinary conditions, according to Cheryl Nickerson and her colleagues at Tulane University in New Orleans, LA" (Matin and Lynch, 2005). This was not the only aggressive finding in this study. E. coli form biofilms much more readily under SMG conditions than E. coli grown on earth at ordinary gravity. "Biofilms in space pose an additional peril because the microbes within them tend to be highly resistant to antimicrobial agents, meaning they are yet another threat to the health of astronauts" (Matin and Lynch, 2005). In addition to microorganisms becoming more deadly they also become more resistant to stresses. Stationary-phase cells were under SMG for 24 hours and they became super-resistant. This becomes an issue when looking at human beings going into space due to the fact that they are immunocompromised and these bacteria are a danger to them during prolonged missions (Matin and Lynch, 2005). If something goes wrong and dangerous pathogens find their way on to the spacecraft then the health of the people on the mission will be put at a very high risk.

There are a multitude of physical and psychological problems that can occur in people partaking in space travel for long periods of time. Broken bones and renal stones from reabsorbed bone material are very common due to bones losing mass while in space. In addition to weak bones, muscles also weaken and blood production is decreased, doing damage to the cardiovascular system. There are also many issues regarding microorganisms and the way they react with the body. Latent viruses (Varicella zoster for example) tend to reactivate. And astronauts also show an increased susceptibility to infections (Matin and Lynch, 2005). This stresses the importance of screening before the journey. Another example of this can be seen from the example of a "...A traveler from Copenhagen carried measles virus with him to the Faroe Islands in 1846, and 6,000 of the 8,000 inhabitants caught the disease" (Morens, 2014). If an individual unknowingly carries a virus they are putting every other person on the mission at risk by spreading germs. According to research, crew members are actually the most predominant source for bacteria, with ground supplies from earth being the other major source (Gueguinou et al., 2009). All risks are heightened with pathogens when air, food, water, and waste are all recycled in a confined area for long periods of time. This confinement helps with the microorganisms being transferred among the crew and having disease and illnesses spread throughout the spacecraft (Gueguinou et al., 2009). The close quarters that people are going to be in will attribute to the spread of bacteria that is present.

Due to these issues that are presented for the members of this mission it is important to know that none of the individuals going aboard the spacecraft have any type of dormant virus. It is also necessary to have proper precautions for diseases that may break out over the course of the journey. Whether that is having a large stock of antibiotics and medications on board or having the supplies to manufacture many types of medicines while on the way to the new planet. If proper precautions are taken it is very possible that the issues individuals will have to deal with medically will be minimal. One of the most important things to look at other than protection from germs would be protection from individuals who pose a threat to others; whether this threat be from, somehow contracting a disease, mental illness, or breaking laws.

When looking at protecting the individuals on the spacecraft from disease it is important to first look at and understand how diseases and illnesses will spread in the environment. The density of a population is a major contributor when looking at the spread of disease. Based off of studies on Earth (since there is not yet experience in space) issues with disease become more prevalent when the area is more populated. When a population tends to become overcrowded it often leads sanitation issues. Sanitation issues cause the rate of agent transmission to be very high in populated areas (Morens, 2014). It is important to focus on the population of our mission to ensure that over population does not occur and lead to issues regarding sanitation and spread of disease. If the living spaces become overcrowded then they can potentially become breeding grounds for infectious agents. The way the sanitation issues and spreading of diseases occurs is the constant recycling of resources between humans as well as other hosts that carry similar diseases (Morens, 2014). It is important to monitor the population and keep it steady as to prevent overcrowding within the spacecraft.

In addition to overcrowding, changing eating habits can also affect disease. When an individual is given a new diet then the body and microbes all react differently and can lead to issues regarding viruses and disease. This is where food poisoning can come from and when multiple people are exposed to the outcome of food poisoning more things can spread to the larger population (Morens, 2014). Making sure that the resources are intact to keep a steady diet is vital in being sure that this aspect of disease spreading can be avoided.

One more thing that greatly contributes to the spread of infectious microbes is what certain individuals are exposed to. Children often have poorer hygiene then adults and therefore can be a breeding ground for infections. Colds, coughs, sore throats, etc. are often passed around a day care because children don't know how to properly protect themselves from the germs. A breakdown in hygiene can cause an entire class to be affected by these issues (Morens, 2014). Besides children, what people are exposed to in their line of work is also of concern. In this community people are going to be living in close quarters and interacting with others on a daily basis. People such as agricultural workers or garbage men, while extremely vital to the community, are exposed to things that lawyers and engineers are not. Farmers are often exposed to damp conditions. Damp conditions are where disease microorganisms tend to flourish (Morens, 2014). People in these fields need to be monitored that they are taking all necessary precautions to prevent contracting an illness and therefore prevent passing an illness on.

In addition to looking at infectious diseases and how to protect people from such occurrences it is also important to look at the issue of mental illness and when individuals become a violent threat to others on board. The belief that all people with any type of mental illness are dangerous is extremely false. In most cases people are of no danger whatsoever to the people around them. However, this does not mean that violent individuals do not arise. If individuals become violent, even if it is not due to mental illness, it is of vital importance to protect the rest of the community from them. One study showed that the most effective way to prevent attacks and protect individuals is to quarantine the violent parties. The main question at hand was whether or not "civil commitment, as practiced by community institutions, reduces violent crime" (Catalano and McConnell, 1996). The findings implied that committing people meant fewer crimes for the community. Individuals that were placed in quarantine would have often committed a violent crime if left in the community (Catalano and McConnell, 1996). This proves that isolation and quarantine is a viable option for violent individuals.

The issue that then arises from this conclusion is whether or not the spacecraft will have the capacity and resources to quarantine the individuals that become violent. Quarantine is a very plausible solution until resources begin to become depleted.

The method to employ to ensure that people that are a threat to the mission are not depleting resources is execution. If someone is such a detriment to the mission and has harmed other passengers then they should be removed from quarantine and executed to prevent them from depleting resources as well as to prevent them from further harming others. The most plausible way for this to occur would be to eject them from the spacecraft into space. Prior to expelling the individuals from the spacecraft they would be induced with anesthesia. According to Mosby's Medical Dictionary, after anesthesia is administered unconsciousness, muscle relaxation, and the absence of feeling pain will ensue (Mosby's Medical Dictionary, 2009). Once the anesthesia has been administered, either by inhalation or IV injection, the individuals will not be conscious and would not suffer too terribly when set into space. When put into space without the proper protection one would lose consciousness after approximately 15 seconds (Starr, 2014). If the individual is administered anesthesia prior to being released in space then their death will be relatively painless due to the fact that they will already be unconscious and will not have to suffer for those 15 seconds waiting to lose consciousness. The person would eventually die due to the lack of oxygen in outer space (Starr, 2014). This same method can be used if it is decided that extremely elderly people are no longer a use to the mission and they are simply depleting resources. Opposed to other options, this would be the most ethical way to ensure that people are not feeling a lot of pain and suffering during the process. Once the individuals have perished, it would be beneficial to keep them connected to the spacecraft in order for their bodies to be used as biodegradable material for fertilizer.

It is of vital importance that individuals on this mission are aware of how to handle dangerous situations.

There will be medical professionals on board that will be in charge of things regarding disease and bacteria discussed above as well as administering anesthesia to individuals. If people know how to handle situations that potentially may cause a risk to patrons then they can act proactively and effectively stop many issues before they become dangerous to the population.

5.5 REFERENCES

Bamford, R., Todd, T., Gibson, K. J., Stamper, R., Norberg, C., Hapgood, M., et al. (2008). "The interaction of a flowing plasma with a dipole magnetic field: measurements and modeling of a diamagnetic cavity relevant to spacecraft protection." Plasma Physics and Controlled Fusion, 50(12), 124025.

Catalano, R. A., & McConnell, W. (1996). A Time-Series Test of the Quarantine Theory of Involuntary Commitment. Journal Of Health & Social Behavior, 37(4), 381-387.

Collin, A., Lechene, S., Boulet, P., & Parent, G. (2010). "Water Mist and Radiation Interactions: Application to a Water Curtain Used as a Radiative Shield. Numerical Heat Transfer, Part A: Applications" 57(8), 537-553.

Deblois, B.M., Garwin, R.L. & Kemp, R. (2005) Star Crossed. IEEE Spectrum, 42(3), 41-49.

Downs, Raymond L. Less lethal weapons: a technologist's perspective. (2007). Policing, 30(3), 358-384.

Evlanov, E. E., Zavjalov, M. M., & Tyuryukanov, P. P. (2013). Electron guns for spacecraft. Cosmic Research, 51(5), 388-395. doi:10.1134/S0010952513050043

Gargaté, L., Bingham, R., Fonseca, R. A., Bamford, R., Thornton, A., Gibson, K., et al. (2008). "Hybrid simulations of mini-magnetospheres in the laboratory." Plasma Physics and Controlled Fusion, 50(7), 074017.

Gueguinou, N., Huin-Schohn, C., Bascove, M., Bueb, J. L., Tschirhart, E., Legrand-Frossi, C., & Frippiat, J. P. (2009). Could spaceflight-associated immune system weakening preclude the expansion of human presence beyond Earth's orbit?. Journal of leukocyte biology, 86(5), 1027-1038.

Kaurin, P. (2010). With Fear and Trembling: An Ethical Framework for Non-Lethal Weapons. Journal Of Military Ethics, 9(1), 100-114.

Matin, A., & Lynch, S. V. (2005). Investigating the threat of bacteria grown in space. ASM News, 71, 235-240.

Mehlhorn, T. (2014). National Security Research in Plasma Physics and Pulsed Power: Past, Present, and Future. IEEE Transactions On Plasma Science, 42(5), 1088-1117.

Morens, D. (2014, April 11). Population density. Retrieved October 30, 2014, from http://www.britannica.com/EBchecked/topic/287492/infectious-disease/12989/Population-density

Mosby's Medical Dictionary, 8th edition (2009).

Rappert, B., & Wright, S. (2000). A Flexible Response? Assessing Non-lethal Weapons. Technology Analysis & Strategic Management, 12(4), 477-492.

Schimmerling, Walter. "The Space Radiation Environment: An Introduction". The Health Risks of Extraterrestrial Environments. Universities Space Research Association Division of Space Life Sciences. 12/05/2011.

Schneider, D. (2005). To Boldly Go (Again). American Scientist, 93(4), 312-313.

Seed, Thomas. "Acute Effects". The Health Effects of Extraterrestrial Environments. Universities Space

Research Association, Division of Space Life
Sciences. 5 December 2011.

Starr, M. (2014, July 27). What happens to the unprotected
human body in space? - CNET. Retrieved
November 3, 2014, from
http://www.cnet.com/news/what-happens-to-the-
unprotected-human-body-in-space/

Weinberg, H. S. et al. "Very High Mutation Rate in
Offspring of Chernobyl Accident Liquidators."
Proceedings of the Royal Society B: Biological
Sciences 268.1471 (2001): 1001–1005.

Volume of Revolved Trap V_{RT}

$\Delta V = \pi y^2 \Delta x$

$y = mx + b = \frac{1}{2}x + 42.25$

$y^2 = \frac{1}{4}x^2 + 42.25x + 1785$

$V = \int_0^{85.3} \pi \, (0.25x^2 + 42.25x + 1785)\,dx$

$= \pi \left[\int_0^{85.3} 0.25x^2\,dx + \int_0^{85.3} 42.25x\,dx + \int_0^{85.3} 1785\,dx \right]$

$\quad\quad = \quad\quad\quad\quad = \quad\quad\quad\quad =$

$\left.\frac{.25x^3}{3}\right|_0^{85.3} \quad \left.\frac{42.25x^2}{2}\right|_0^{85.3} \quad \left.1785x\right|_0^{85.3}$

$51,721 \quad\quad 153,707 \quad\quad 152,261$

$\pi \, 357,689 = 1,123,712 \ m^3$

Volume of Revolved Lower Part V_{LP}

$V = \int \pi y^2 dx \quad y^2 = 8x$

$V = \int_0^{25.35} \pi \, 8x \, dx = \pi \cdot \left.\frac{8}{2}x^2\right|_0^{25.35} = 8,076 \ m^3$

Volume of Revolved Upper Part V_{UP}

$V = \int_0^{6.9} \pi 4x\,dx = \pi \cdot \left.\frac{4x^2}{2}\right|_0^{6.9} = 1,795 \ m^3$

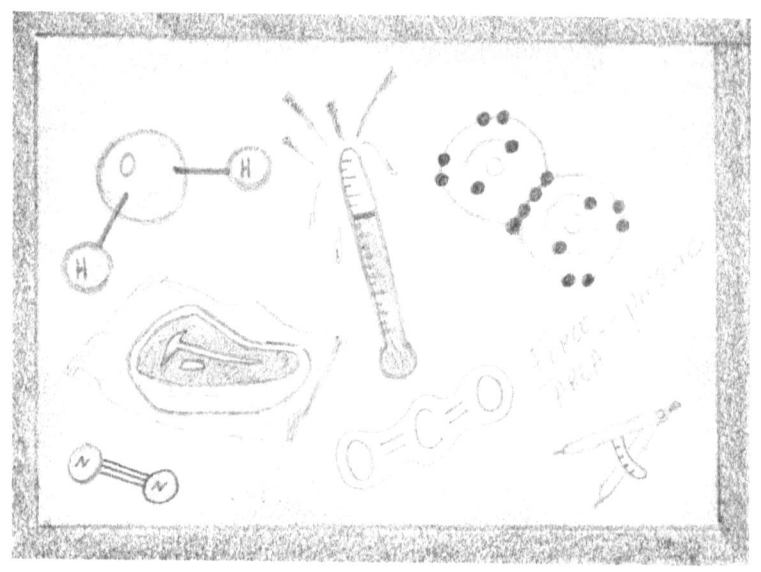

Chapter 6: Resource Management
By Katie Askegaard, Tim Betts and Kara Elser

6.1 INTRODUCTION

Properly managing spacecraft resources on the flight to Kepler 186f is a vital component for the success of this mission. Throughout the research conducted, we have addressed the major questions regarding how to make the spaceship sustainable to last 200 generations. We focused on managing resources detrimental to the survival of the crew, such as food and water, and resources detrimental to the functioning of the spaceship, such as climate control and waste management.

Having food is crucial for crew survival, and crops will be grown through a hydroponic system that uses a nutrient dense water solution for crop vitality. It is essential to have a water filtration system, therefore we decided on reverse osmosis. For ship cleanliness, we will manage waste by using wheatgrass and thermal composting. All these functions will be supported by a comfortable environment with plants regulating oxygen and CO_2 levels, a biofiltration system to control trace gases, and an air ventilation system to circulate the air and control for temperature.

Addressing the education and medical resources are another vital component in mission success to keep the crew healthy and able to perform all the roles required for the ship to function. For education, we will have an i-cloud video system, and for medical resources, we will have primary health care supplies. In regards to maximum capacity and logistics, we will have enough resources to sustain a 450 person population on the space craft and we will have a resource management team to ensure equal distribution of resources.

6.2 SUSTAINABLE FOOD PROVISIONS

Making food sustainable is of the utmost importance for this upcoming mission. Supporting approximately 450 people for 200 generations will require a closed loop bio regenerative system to provide the people with adequate nutrition to sustain the population. If the ship's food supply is unsustainable, the mission will be futile. Having a closed loop food supply system with crops not only feeds the crew, but it also regulates the environment with CO_2 and oxygen control. Thus, having a stable and efficient food supply system is vital for the habitability of the spacecraft.

Experimentation for deciding what kind of plant growth system to use in space is a heavily researched topic in space science because having pre-made food is unsustainable on long-term missions. Soil clearly yields sufficient crop levels, as seen on earth, and is one option for the mission to Kepler 186f (Nelson et al 2013, Nelson & Wolverton 2011, Nelson et al 2007). The second option is a hydroponic system, which is a soilless system where crops use a nutrient-dense water treatment to grow (Nelson et al 2013, Paradiso et al 2012). Both soil and hydroponic systems are viable options for the mission and can yield a wide variety of crops.

The success and sustainability of food production in a soil system is evident from how crops are grown here on Earth, but it has additional benefits if used in space. First and foremost, soil microbes can metabolize trace gases to purify the air (Nelson et al 2013, 90). Reducing harmful gases in a closed spaceship environment is of the utmost importance for crew vitality, making this a major benefit of a soil-based system. Another benefit is the ability to recycle crop and human waste into a soil system (Nelson et al 2007, 677). Space is limited and waste must be recycled or eliminated, so using waste to fuel a soil system is an efficient option. Finally, a soil-based system opens up the possibility of creating agriculture on Kepler 186f by blending our soil with the foreign soil (Nelson et al 2007, 677). If settlement is indeed possible, having agriculture on the foreign planet is important for creating a sustainable colony.

A soil system has drawbacks as well. Firstly, there is the potential for disease in the soil given that waste is recycled into the system (Rius-Ruiz et al. 2014 92). Also, while soil works well on Earth, the low gravity conditions of a spacecraft provide different growing conditions. How the roots will grow, how the soil will stay compact, and how water will effectively nourish the plants are all factors that could limit the success of plant production. Soil-based agriculture will also take up a significant amount of space to successfully nourish 450 people, which is important to consider as well.

A hydroponic system, on the other hand, uses minimal amounts of space because plants can be grown vertically. Not only is this soilless system space saving, but it is also lighter and produces ten times more food than soil methods and saves five to ten times the water (Rius-Ruiz et al., 2014 92). While hydroponics is not as common for growing food on Earth, experimentation shows that this system produces a high crop yield with good seed quality (Paradiso et al 2012). The risk of disease is also reduced as long as the nutrient solution is recalculated and proper system sanitation procedures are followed (Paradiso et al., 2012 1502).

A drawback of a hydroponic system is that it takes away the possibility of combining soil with the soil of Kepler 186f to facilitate agriculture for colonization (Nelson et al 2013). It also means another way to recycle waste and another way to reduce trace gases is required for ship sustainability (Nelson et al 2013). Another drawback is the fact that a hydroponic system requires more specialized knowledge to successfully operate. Specific ion concentrations and nutrient levels must be determined to have a successful yield, and the availability of these nutrients (zinc, magnesium, etc.) must be sustainable (Rius-Ruiz et al., 2014 92)

When considering the factors above, a hydroponic system will be the most efficient food producer for this mission. First and foremost, the space saving and high yields are of the utmost importance. The NASA standard for caloric intake uses the following formulas for basal energy expenditure (BEE): "For women, BEE = 655 + (9.6 x W) + (1.7 x H) - (4.7 x A), and for men, BEE = 66 + (13.7 x W) + (5 x H) - (6.8 x A), where W = weight in kilograms, H = height in centimeters, and A = age in years." (Dismukes) Thus, a substantial amount of crop yield is required, and the amount of space, water, and soil needed to adequately meet these demands if a soil-system is used is unachievable. The ability to maintain plant nutrients in an exact manner in a hydroponic system is also a vital reason for choosing this system because it leads to growing crops in the most efficient, resource-saving manner.

However, a small soil growth area should be present on the spacecraft as well in order to help control trace gases and to provide the soil needed for colonization. This plot will produce some and will be primarily used to reduce trace gases. It will be discussed in more detail later.

The first step for implementing this system is to test and develop the most efficient design for the vertical structures the crops will be grown on. Based on evidence from Biosphere 2, a variety of crop seeds are needed, including rice, wheat, sweet potato, millet, beans, and potato among others (Nelson et al., 2007). LED lights will be used to provide the energy needed for plant growth because of their small mass and energy efficiency. Chlorophyll and carotenoids respond to the light produced by blue and red LEDS, making it an efficient energy source (Yeh, Chung 2009, 2178). The final step for implementation of the food source is to have a wide variety of crops at different stages of growth in order to keep the interior spaceship environment stable (Drysdale 2003, 54).

6.3   ENVIRONMENTAL CONTROL AND LIFE SUPPORT

Controlling the spaceship environment to have livable conditions is of the utmost importance for crew survival and comfort levels. Components to control include having a livable temperature, proper CO_2 and oxygen levels, airflow, and minimal trace gas build-up. The ship cannot rely on oxygen storage or short-term solutions; a closed-loop system must be installed to support 450 people for roughly 200 generations.

One potential solution is to have an Environmental Control & Life Support System (ECLSS) that uses technology to regulate the environment. Jones and Kliss propose a recycling system seen in figure 1 that is 90% efficient in recycling oxygen and water and that controls for trace gases as well as waste management (2009, 921). Samsonov et al proposes another system (see figure 2) that also controls for trace gases, maintains oxygen and CO_2 levels, and manages for waste (2000, 131).

Rather than using a purely technological system to regulate the environment, another option is to allow the plants to regulate the CO_2 and oxygen levels with the support of ventilation for temperature control and air circulation. Xiangli Li proposes a ventilation system that is "crucially essential to achieve the exchange of matter and energy in cabin under the microgravity conditions (2012, 203)." In this system, the volume of air in the condenser will control temperatures, and fans with a refrigerant dryer will be used to ventilate the cabins (Li 2012, 204). While Li looks to ECLSS systems, such as the ones previously discussed, to control trace gases, soil biofiltration can also be used (Nelson & Wolverton 2011, 583). Putting an air pump under soil replete with microbes removes a large percent of pollutants. This technology dates back to 1900 Germany when soil biofiltration was invented to control for pollution in the air from factories (Nelson & Wolverton 2011, 585). This system was also used in Biosphere 2 to control for harmful trace gases (Nelson et al. 2013).

The positive of an ECLSS is that atmospheric conditions can be easily calculated and regulated. The modeled systems seen in figure 1 and figure 2 create one, coherent, efficient system that sustains itself. However, this type of technology is yet to be proven in flight and the technology is not yet mature (Jones & Kliss 2009, 925). Problems include "high power consumption, difficult maintainability and logistics, sensitivity of several components to particulates and fouling, gravity related problems in multi-phase fluid flow and separations, and the lack of fine particle settling in microgravity (Jones & Kliss 2009 926)." These problems raise the question of whether or not this system is mature enough to last for 200 generations.

Soil biofiltration has been proven to be effective at controlling trace gases in Biosphere 2 (Nelson & Wolverton 2011, 585, Nelson et al. 2007, 680). This system also naturally responds to changes in pollutants and new types of unforeseeable trace gases (Nelson & Wolverton 2011, 589). Minimal human intervention and regulation are needed to maintain a soil biofiltration system, which is another positive over a complex ECLSS system. Such a system has multiple purposes because it can be used to grow crops (Nelson et al., 2013).

There are potential health problem posed by this system because wastewater could carry human pathogens of infectious diseases that could become airborne (Nelson & Wolverton 2011, 587). Relying on plants to maintain an adequate C02 and oxygen balance is risky as well because proper crop rotation must be implemented in order to avoid having soil and human respiration exceed photosynthesis (Nelson et al 2008, 679).

A ventilation system can properly manage airflow by mathematically managing the levels of air volume, temperature, and pressure to maintain a stable environment (Li 2012, 205). While this system has not been flight proven, the positive effects can be seen in closed-in office buildings. However, this ventilation model is complex, and requires skilled engineers to maintain the system (Li 2012, 217).

Considering the arguments above, a soil biofiltration system with an air ventilation system will be used on the mission to Kepler 186f. With current technology, an ECLSS system is simply too underdeveloped and unreliable. Soil biofiltration, on the other hand, has been proven to successfully reduce trace gases and adapt to changes in the environment. This soil system will not waste space because it can also be used to harvest crops. While hydroponics will be the main source of food supply, soil biofiltration can provide a back-up in the events of hydroponics failing. In the event of colonization, soil can also be integrated into the host planet's soil.

In regards to disease risk with soil, the health team can be informed of this potential problem to ensure that crew members with infectious diseases will be quarantined and that their waste will not be recycled. Addressing the issue of having balanced oxygen and C02 levels, rotating the growth of the crops can mitigate the potential problem (Nelson et al 2008, 679).

It has been decided to use a ventilation system because it is the only way to circulate the air and maintain a comfortable temperature. Plants will be grown in a separate chamber, so the oxygen must be able to flow throughout the spacecraft for the crew to survive.

The first step in implementing the ventilation system is to calculate the optimal parameters for pipe size, pipe flow, and fan pressure. Calculations performed in Li's research are for a spacecraft much smaller in size; thus, adjustments must be made. This system must be installed as the spacecraft is being built and it must be designed so that it reaches each chamber of the ship.

The soil biofiltration system can be installed following the completion of spacecraft construction. This, along with the other hydroponic plants that regulate CO_2 levels, should be functioning prior to take off to achieve proper crop rotation to adequately maintain the environment. Back-up precautions should be installed in the ship as well. A CO_2 scrubber and an oxygen generator should be available in the case of an emergency, and it is recommended that a catalytic converter also be present in the event of excessive trace gas build-up.

6.4   MEDICAL AND EDUCATIONAL RESOURCES

While on the spacecraft, the multi-generational mission will need medical and educational resources to survive. These resources are essential for the mission to survive. In current research in the field of education, most educational resources have been moving toward online videos and seminars. This mission is going to take thousands of years and multiple generations to reach Kepler 186f. In saving space and resources, it will be beneficial for the mission to have interactive educational videos. These interactive videos can serve as recreational as well. It will be able to teach students and children of jobs or other work that needs to be completed while on the mission. This way the population on the ship will not be bored, but also learning something at the same time. There will be multiple videos on different aspects of education. There would be videos about every possible job that will be on the ship.

Learning information about an occupation or job can be frustrating or confusing. However when you have an educational video, individuals can learn at their own pace. That way these educational videos can include instructions on how to complete jobs or ways to use equipment. The members on the ship can also have an iCloud system where they store all of these educational videos. In this place the leadership on the spacecraft can edit the information, especially when it comes to equipment. For example nurses have to know how to use equipment and technology that changes constantly. These videos or website can have these educational resources, and keep up with the advancement of technology on and off the spacecraft (AORN 2012). These websites of videos will be set up where the individuals who need the information will have access to it at any time. These sources will consist of facts and videos of how the equipment will work (AORN 2012). The information presented or created will be monitored and approved before viewing or posting on the cloud system (AORN 2012). While having educational videos meant for grown individuals, we can have other forms of education, while also providing entertainment on the spacecraft.

Educational entertainment will help with the survival of generations on this spaceship. Individuals on the ship need some form of entertainment, especially the younger individuals. This educational entertainment can be in the form of a video game. The individuals looking to have some fun during the day can start by playing these games. In these games the storytelling can be educational, motivating the students (Computers in Human Behavior 2014). The story telling will be appropriate for the age group of 3 to 7 years of age (Computer in Human Behavior 2014). These video games will allow these young individuals to follow through the game learning something from it. Whether it be the alphabet or sentence structure or how to use certain equipment, individuals will be able to learn something new, while also enjoying their entertainment. This educational and recreational component integrate together in a narrative model, producing something individuals on the ship can use (Computers in Human Behavior 2014). The video games can also be created for older individuals and made accessible to other who would like to enjoy the entertainment. Thus producing an inclusive fun environment on the spacecraft.

Creating these educational videos and video games is crucial to the success of the mission. It is important that individuals know how to do their job while on the mission. If people on the craft don't know have to help sick individuals or work certain equipment, people will not survive. It is necessary we provide this system because it doesn't require much space on the craft. It completes the job of educating, but isn't taking up the limited amount of space that is available on the ship. While also providing entertainment to individuals who want to have fun, yet they are learning something new. However medical resources are also needed on the ship.

Medical resources that are provided on the ship are necessary to the success of the mission. It is essential that the proper resources are brought onto the ship for the mission. On the ship and during the mission, we are only going to provide the bare minimum of primary healthcare. It will take up to much space if the missions require something different. The equipment that is going to be brought onto the spacecraft is the basics (ECHO 2001). There are multiple aspects to the equipment needed; therefore some of the equipment that will be brought on the ship is 'basic' medical products. Basically all of the common supplies doctors use to provide a safe and sanitary environment. Also other equipment such as blood work machines will be brought to help control sicknesses and other necessities like that (ECHO 2001). Overall, medical supplies are extremely necessary in this mission to ensure it will be successful. Therefore, it is important that the medical supplies on the ship are good to use and the individuals in charge know how to use the products. Thus, educational and medical resources are important for the success of the mission and should be a priority for the multi-generational mission.

In conclusion, it is essential that individuals on the ship have access to educational and medical resources. These two resources are extremely important in the success of this mission. Without these resources, individuals on the mission would not be able to survive. That's why the education will be made into videos and video games accessible via a cloud system, and only taking the necessary medical resources to save space on this spacecraft.

It is also important, for the success of the mission, that the spacecraft can hold everything that the mission will need. There needs to be enough room on the spacecraft to hold everything, however space is very limited. It is important that individuals in charge of the mission use all the space wisely. It is important that everywhere there is room, we make sure something can fit in that. For example we need to make sure the waste system in place is able to fit. We can use idea such as ones from other countries such as Japan. We can take their idea of eco-towns and create a smaller form to fit on the ship (Cleaner Production 2012). We can do this by creating trashcans into the walls of the spacecraft and have everything enter into one space. Also creating a recycle option making the ship able to do something with current materials already on the ship. That way the mission is maximizing the space available on the ship and using it to better the mission.

Using all available space is important, but so is have a resource management team. It is important that resource management on the ship has a leadership in place that will be able to handle and control the whole department. A system that the department can adapt is the Integrated Resource Management model (GeoJournal 1991). This model is used to make sure the system is set up correctly, and works properly. It is a system where there is detailed planning phase and communication to ensure resource success (GeoJournal 1991). This model can be used on the ship to set up the system in which the department works and functions. This will ensure success of resource management and the success of the ship. However for max capacity and other limitations like that, it will depend on how big the spacecraft is, but the expected size is 450 individuals. In these, resources has to make sure there is enough space and resources, such as food, for this population. Therefore, resources has to work closely with science to make sure all empty space is being used properly. In conclusion, if resource management makes sure it uses the available space on the ship wisely and that management is in sync, the mission will be a success.

6.5   SANITATION AND WASTE DISPOSAL RESOURCES

One of the critical questions for this mission is how to make human waste sustainable on the spacecraft. The issue of waste is not a topic to be taken lightly on a multigenerational spacecraft that has approximately 450 humans aboard for both sanitary and practical reasons. It's neither healthy to live in close proximity to human waste due to diseases this could cause nor is it practical to store human waste (or any waste) for thousands of years. There is a wide array of options of how to do this including composting toilets and wheat grass with everything in between.

The first possible option for how to make human waste sustainable on the space ship would be by using composting toilets. Ellen Rowland offers the idea of using a composting toilet which would be an attainable way to use our resources in a closed-loop process to create a nutrient-rich fertilizer which is needed to grow plants on the ship (Ellen Rowland 2013). With an analysis of the pros and cons of the dry toilet, it will be determined if this would be what is best for the space ship.

There are two benefits of the composting toilet. It would preserve water that would be valuable on the ship because new water would not be easily supplied from Earth as well as provide a nutrient-rich fertilizer. The fertilizer would enhance the ability to grow food on the spaceship which is also detrimental to the future of the ship. Due to our lack of contact with a habitable world for several generations, the ship needs to be operated in a closed-loop process (Rowland 2013). There would only need to be items for a natural covering aboard the ship such as straw, sawdust, peat moss, coconut core, peanut shells, etc. to use in the composting toilet (Rowland 2013). This would be feasible because these remains would not be used elsewhere on the ship. Other benefits of a composting toile include: controlled odor when properly maintained, conserves water, reduces energy consumption and greenhouse gas emissions (Rowland 2013).

There are some downsides of the composting toilet. First and foremost, there must be a section of the toilet that can be removed and emptied into the outdoor compost. The outdoor compost is where garden clippings, egg shells, fruit peels, etc. will be added to enhance the nutrients in the compost (Rowland 2013). This poses a problem because there is nowhere "outdoors" to put the compost because it is just one space ship. If there was a way to ameliorate this problem though, it could be feasible because it would be a great way to fertilize, use excess food on the ship that humans don't eat, and get rid of human waste.

Thermal composting toilets offer another way to get rid of human waste. Researchers conducted a study to analyze different types of waste from toilets and how to most efficiently disinfect these wastes. The study was completed on both large scale and small scale studies. It is very important to disinfect the feces because they contain 10^{10} microorganisms per gram dry matter and some of these microorganisms are pathogenic. Without disinfecting the feces, it cannot be used for fertilizer or soil conditioner. A more reliable way to disinfect the feces is through heat inactivation which is also known as pasteurization. There are two ways to use heat activation, either incineration or by aerobic digestion. In order to have heat inactivation, the vessel must be insulated to retain the heat (Vinneras et al. 2003). There are several benefits of thermal composting toilets, but there may pose some difficulties for using this method on the ship, too.

The most important part of human waste management is achieving low risk for contamination on this space shuttle. The pilot-scale experiment showed high safety margin of total inactivation with composting faces and food waste. This is a promising sign for thermal composting and using the compost as a fertilizer that has low risk of spreading pathogens on the ship (Vinneras et al. 2003). This presents promising research for using the thermal composting toilet on board the space mission. It allows the inhabitants to have a fertilizer. However, there may still be some dangers such as it is not fully perfect. Although, the only way to have no risk for contamination on this ship is to never use the feces in another way. This is not possible because the ship needs to use a close-looped process or they will lose too many nutrients. Another danger is having a portion of the space craft being extremely hot to ensure the processes take place and it is a time-consuming process.

Sometimes, it is even simpler to use a natural way to have a sustainable disposal system. Sciences have created a fully sustainable disposal system by using the byproduct of wheat. This has been used in the talks of creating a manned mission to Mars.

The wheat grass is converted in to carbon while heated and the gases from waste incineration are sent through said carbon. The carbon absorbs the nitrogen oxide and converts it into nitrogen, ammonia, and nitrates (American Society of Agricultural Engineers 2003). Now it is important to see if wheat grass would be plausible on the ship and why it may be useful for this long-term mission.

The benefits of using wheatgrass are it is the inedible part of the wheat plant and it can reclaim pollutants produced while burning waste on the spaceship. It is an excellent resource to use because otherwise it would be thrown away on the ship (creating more waste) but now it could be used for something more useful. Another benefit of this is it can be used to replace cabin pressure leakage and be used as a fertilizer (American Society of Agricultural Engineers 2003). The only concern now is it the space mission will be growing wheat on board. It is probably the most beneficial thing possible because it can be eaten and can be grown hydroponically. It is essential that there is a nutrient solution where it can be grown and where it will be exposed to artificial sunlight (American Society of Agricultural Engineers 2003). This is perfect for our space shuttle which is developing several ways to create nutrient rich fertilizer through waste management and the space shuttle relies on mimicking Earth-like conditions aboard the ship which would include artificial sunlight.

After a comparison of the positive and negative aspects of each of these waste management solutions, it would be best to combine two of these ideas. These two ideas are thermal composting and wheat grass. The thermal composting is selected because of the hygienic safety as well as the more successful rate of destroying pathogens compared to the regular composting (Vinneras et al. 2003).

The thermal composting also offers a great way to have fertilizer aboard the ship which would be necessary when we grow food to support the inhabitants. Using wheat grass is also a great idea aboard the ship because it is something we would otherwise throw away. Now the wheat grass can be used to absorb the nitrogen oxides when the waste in is incinerated (American Society of Agricultural Engineers 2003). This is very important when the ship will be using thermal composting with incineration. The wheat grass also adds the benefit of alleviating cabin pressure leakage.

These two waste management solutions will be easily implemented on the space ship. Wheat will be grown on the ship to feed the people living on it. It can be grown hydroponically and then the waste management group can use the excess from the wheat in ways listed above. When designing the ship, there will need to be an area reserved for thermal composting. This is important that it is not located close to where people are living to enhance sanitary reasons. It also must be designed to withstand the heat to incinerate the feces. It will be crucial that these are part of the space ship to use the nutrients from feces in a meaningful way but also to not worry about pathogens within the feces in the closed-loop process that is being created.

Another critical question of the mission is what would be the best water filtration system that we could use to ensure that we have an adequate amount of water at all times aboard the ship. For four crewmembers for a life at the space station, it would require 40,000 pounds of water from Earth to be resupplied every year (Barry and Phillips 2000). The mission has over 100 times the amount of people that this reference would be for and they are traveling much further than the space station and an easy resupply of water from Earth would be even less practical. Receiving approximately 4,500,000 pounds of water from Earth every year to sustain the ship is absolutely infeasible.

Therefore, the ship itself needs to act in a closed loop to sustain its own water.

It is essential that the ship finds an efficient way to filtrate our water in space or else it will leave the people on the space shuttle without water because there is no way that we can bring aboard enough water to sustain several hundred generations. There are many types of water that are included before it is treated such as hygiene water, condensate water, and urine which contain ammonia, halogenated carbons, and heavy metals which would be dangerous for humans (Lee and Leuptow 2001). One possible way to do this is through a reverse osmosis filtration system. Reverse Osmosis (RO) removes ions and organic pollutants. RO can be used to recycle wastewater throughout a space mission in a compact process. The study also stresses several of the important design requirements of Waste Recovery Management System (WRMS) such as being reliable, capable, efficient, and having minimal expendables. WRMS also seeks to minimize the weight, volume, power consumption, and cost (Lee and Leuptow). The RO shares many great things in comparison to the WRMS design requirements, but there are still many parts that need to be further looked into.

There are several reasons that shuttles are looking towards RO for long term space missions. These include it being a regenerating technology that does not need to be replaced as often as other conventional filtration systems. The reverse osmosis membrane filtration only needs to be replaced once or twice a year in commercial membrane plants. It has the benefits of removing ions, proteins, and organic chemicals which conventional treatments struggle to remove. Lastly, it is an absolute filtration method which leaves it stable and predictable (Lee and Leuptow 2001). These benefits are exactly what are needed on a long-term space mission like the one the class is designing.

The downside of using the RO is it is much more

complicated than the conventional filtration. Therefore, it would require a fundamental understanding of the chemistry and physics because these explain why some pollutants are rejected by RO. (Lee and Leuptow 2001). The study cites issues such as RO rejection depends on the physical chemistry of the solute, solvent, and membrane and the size difference between the solute and membrane pore (Lee and Leuptow 2001). It goes onto state that there are several of the same problems for ions. This could prove problematic on the space shuttle because these kinds of issues are not always common knowledge.

Although the ship should have gravity and Earth-like conditions, that may not be fully possible. Therefore, it is important to look into water filtration systems that would still work without gravity present. It is also extremely important that the water filtration system is sanitary because there is only one ship; so, the inhabitants would be living very close to this process. This brings forward the possibility of a membrane-aerated, membrane-coupled bioreactor (M2BR). M2BR was designed for long-term space missions specifically for achieving aeration and biomass separation that can survive under microgravity conditions. It does so with a gas-transferring membrane and provides bubble-less gas transfer to the wastewater. This is different than conventional aerobic biological wastewater treatment that relies on gravity to create air bubbles. There are two membranes, one has a biofilm containing nitrifying and denitrifying bacteria and the second one provides the hygienic safety constraint for people living in close proximity to the M2BR (Chen et al. 2008). The hygienic part is important for our space shuttle because the people will be living on the same space shuttle as where the water will be treated.

During the study, they were able to achieve their goal when the total nitrogenous pollutant removal and COD efficiencies were over 90% (Chen et al. 2008). However, it brings up some important questions that if there was gravity present on the space shuttle, if the M2BR would still act similarly. It is optimal to have gravity and Earth-like conditions on the mission because these are not astronauts that are going into space. This provides a great back-up plan if there was no gravity. Secondly, another large benefit on this type of filtration system is that it is hygienic with the second membrane. This is important when the people on the mission will be living in such tight spaces with the filtration system and if the system makes them sick or the filtration system is failing, it would be deadly for the ship. This would render the mission useless.

Reverse Osmosis would be the best option to use as a water filtration system on the space mission. This is for several reasons such as the M2BR relies on the fact that there would be no gravity present. It poses some issues on if that would react the same way in an environment that has gravity. The mission hopes to have the same conditions on Earth including gravity, so it should be assumed that the process used on the ship should take into consideration gravity. Also, the needing for knowledge of chemistry to use RO should not be a problem because the space mission could continue to train people on the ship to be equipped to handle these problems, especially because this mission is multi-generational. Another great part of RO is the fact that is lasts a long time and needs to only be replaced once or twice a year (Lee and Leuptow 2001). Therefore, the space ship will have a reverse osmosis filtration system built prior to leaving Earth. It will be tested for efficiency during the dry run, and if any alterations are needed, they will be established before sending the space mission to Kepler 186f.

6.6 REFERENCES

American Society of Agricultural Engineers, Wheat grass on a mission to Mars. Resource Engineering and Technology for a Sustainable World. 10.12, 4 (2003).

Barry, Patrick L. and Tony Phillips, Water on the Space Station. NASA Science. (2000).

Chen, Ruoyu, Michael Semmens, and Timothy LaPara, Biological treatment of a synthetic space mission wastewater using a membrane-aerated, membrane-coupled bioreactor (M2BR). Journal of Industrial Microbiology & Biotechnology. 35.6, 465-473 (2008).

Björn Vinnerås, Thermal composting of faecal matter as treatment and possible disinfection method – laboratory-scale and pilot-scale studies. Biosource Technology. 88, 47-54 (2003).

Dismukes, Kim. "Food for Space Flight." NASA Human Space Flight. NASA, 07 Apr. 2002. Web. 23 Nov. 2014.

Drysdale, A.e., M.k. Ewert, and A.j. Hanford. "Life Support Approaches for Mars Missions." Advances in Space Research 31.1 (2003): 51-61.

Jones, Harry W., and Mark H. Kliss. "Exploration Life Support Technology Challenges for the Crew Exploration Vehicle and Future Human Missions." Advances in Space Research 45.7 (2009): 917-28.

Jouvet, Pierre-andre; Gregory Ponthiere. Survival, reproduction and congestion: the spaceship problem re-examined. Journal of Bioeconomics. 13, 233-273 (2011).

Lee, Sangho and Richard M. Leuptow, Reverse Osmosis filtration for space mission wastewater: membrane properties and operating conditions. Journal of Membrane Science. 182.1-2, 77-90 (2001)

Li, Xiangli. Design and Optimization of HVAC System of Spacecraft. N.p.: INTECH Open Access, 2012. 203-18.

Lynn O'Dowd Bell, Developing a Perioperative Educational Video Web Site. AORN Journal. 95, 463-473 (2012).

Manjit Kaur, Sarah Hall. Medical supplies and equipment for primary health care A practical resource for procurement and management (ECHO International Health Services Ltd 2001), pp 63-161.

Martin Florin, Gabriel Erhard. Integrated Resource Management. GeoJournal. 25, 109-113 (1991).

Natalia Padilla-Zea, Francisco L. Gutierrez, Jose Rafael Lopez-Arcos, Ana Abad-Arranz, Patricia Paderewski. Modeling storytelling to be used in educational video games. Computers in Human Behavior. 31, 461-474 (2014).

Nelson, M., W.F. Dempster, and J.P. Allen. "Integration of Lessons from Recent Research for "Earth to Mars" Life Support Systems." Advances in Space Research 41.5 (2007): 675-83

Nelson, Mark, William F. Dempster, and John P. Allen. "Key Ecological Challenges for Closed Systems Facilities." Advances in Space Research. 52.1 (2013): 86-96

Nelson, Mark, and B.c. Wolverton. "Plants Soil/wetland Microbes: Food Crop Systems That Also Clean Air and Water." Advances in Space Research 47.4 (2011): 582-90

Paradiso, R., R. Buonomo, V. De Micco, G. Aronne, M. Palermo, G. Barbieri, and S. De Pascale. "Soybean Cultivar Selection for Bioregenerative Life Support Systems (BLSSs) – Hydroponic Cultivation." Advances in Space Research 50.11 (2012): 1501-511.

Rius-Ruiz, F. Xavier, Francisco J. Andrade, Jordi Riu, and

F. Xavier Rius. "Computer-operated Analytical Platform for the Determination of Nutrients in Hydroponic Systems." Food Chemistry 147 (2014): 92-97.

Rowland, Ellen, Anatomy of a composting toilet. Natural Life. 14-16 (2013).

Samsonov, N.m., L.s. Bobe, L.i. Gavrilov, V.m. Novikov, N.s. Farafonov, Ju.i. Grigoriev, E.n. Zaitsev, S.ju. Romanov, A.i. Grogoriev, and Ju.e. Sinjak. "Long-duration Space Mission Regenerative Life Support." Acta Astronautica 47.2-9 (2000): 129-38.

Satoshi Ohnishi, Tsuyoshi Fujita, Xudong Chen, Minoru Fujii. Econometric analysis of the performance of recycling projects in Japanese Eco-Towns. Journal of Cleaner Production. 33, 217-225 (2012).

Yeh, Naichia, and Jen-Ping Chung. "High-brightness LEDs—Energy Efficient Lighting Sources and Their Potential in Indoor Plant Cultivation."Renewable and Sustainable Energy Reviews 13.8 (2009): 2175-180.

Deck A - Emergency
        controls
        & Aux. Computers
Deck B - Computer-
        History
Deck C - Living Quarters
Deck D - Food Prep & Commissary
Deck E - Cargo - O₂ + Envirns
Deck F - Cargo - H₂O + Envirns
Deck G - Cargo
Deck H - Emergency Prov.
Deck I - Emergency Prov.
Deck J,K - Telescope and
        Telemetry Input
Deck L,M - Shuttlecraft
        Hangar & Outside
        Ports

Chapter 7: Sociological Concerns
By Tyler Durkee and Amy Wynant

7.1 INTRODUCTION

The objective of this chapter is to address the sociological concerns in a long duration mission to the planet known as Kepler 186f.

The mission requires a functioning and efficient crew. Our research aims to provide potential methods to simulate a terrestrial climate that encourages maximum efficiency in an extraterrestrial environment.

The first question we wanted to look at was the ethical guidelines that would be implemented for the mission. We were unable to draw from past missions because no previous mission has ever been multigenerational so we came up with 9 principals from the Institute of Medicine, with the top one being a balance of risk-benefit, which would act as guidelines.

In addition, it is imperative that we set forth a plan to recruit qualified and healthy crew members. Once on the spacecraft, mental health and physical health facilities will need to be accessible.

This leads to the question of what occupations are needed to be self-sustaining. Using research of communes, it was found that there needs to be a diversity if jobs to be efficient and meet all the needs of the mission.

Another question that relates to occupations is how the edication system will work. Using research from studies of education systems it was found that an apprenticeship program would be the most efficient.

Every person on the ship will be necessary, and it will be important to keep the population at a steady level that can be provided for. This can be done through a mix of social pressures and science. Along with this, it will be important to conserve resources and make sure supplies are not wasted on dependent groups that emerge, such as the old, sick, and disabled.

Our final topic of discussion was the relationship and knowledge base passengers would have with Earth and we soon realized the importance of the group mentality. We decided that lying to the passengers would not be in their best interests seeing as this is and Earth-based mission and there would be some communication with Earth. However, we did not want to depress those aboard the mission so we decided to implement Earth-like characteristics such as hobbies and activities to create a happier and more productive society.

7.2   MULTIGENERATIONAL MISSION SOCIOLOGICAL CONCERNS

While sociological concerns may not be the chief issue that needs to be addressed when planning a multigenerational mission, they quickly present themselves in many facets. Due to the complex nature of a multigenerational mission, it can be assumed that change will be one of the only constants throughout. With that in mind, we contemplated how we would maintain ethical standards during the course of the mission. The safety of those aboard a mission is always of top concern for institutions like NASA and we share those interests as well. While there are many sociological issues to address, we felt that figuring out how to maintain ethical standards during the mission would provide guidance and answers to many other questions. In short, we chose this issue because one of the first things we realized was that a mission of this caliber had never been done before and therefore we could not rely on former missions to provide us with this answer. We soon gathered that we were charged with the dilemma of figuring out if this mission was going to produce results that would justify all of the unknowns that present themselves in regards to a multigenerational mission. We realized that while these humans would be completely surrendering their rights and freedoms to the mission, we still needed to prove and take steps to ensure that this mission would be ethical so that we could get this mission funded and approved.

Paul Gilster, space technology expert and writer of the popular website "Centauri Dreams", poses the important question we asked ourselves at the start of this mission, "Do we have the right to subject human beings to multigenerational conditions and raised without true parents in a potentially hostile environment?" (Gilster 2014). This question was our starting point and Dr. Edward Regis Jr., an American philosopher who focuses on science and intelligence, concluded that there is no fundamental ethical distinction between raising children aboard such a ship and raising them aboard our own planet (Gilster 2014). We agreed with Dr. Regis and began our search for ethical guidelines.

First and foremost, people want to feel safe on this mission and in order to fund the mission we need to prove it will be safe and ethical. While there are known and unknown risks in any space mission, all of which can jeopardize both the explorers and the mission, this particular mission aims to push the boundaries further then they have ever been pushed before. However, this does not mean we need to abandon all ethical standards. Space missions are never given a green light if they do not meet certain health and safety standards and this mission should be no different. There are many health risks associated with long duration missions and UV radiation exposure is especially dangerous for the severe risk it poses. An editorial from Space.com written by the staff writer Miriam Kramer suggests NASA make some exceptions to some of their standards in order to allow for such missions seeing as the levels of radiation expose currently violates one of their health limitations (Kramer 2014). While this source is only speaking about a mission to Mars, which would not require a multigenerational trip, this would be the longest mission known to date. All in all Kramer brings up the health aspect of ethics and speaks to violations of the physical variety, things she says need to be coupled with other standards of the ethical variety.

During our search for ethical guidelines, we found that the National Aeronautics and Space Administration (NASA) asked the Institute of Medicine (IOM) to develop an ethics framework and to identify principles to guide decision-making about health standards for long duration and exploration class missions. IOM found that a successful approach to avoiding the need for a single ethical theory is to focus on mid-level principles rather than the theory to which they belong or from which they are derived (Kahn, Jeffery, Catharyn Liverman, and Margaret McCoy 2014). This approach has proven to be particularly successful when used by expert committees or commissions made up of individuals with diverse commitments in an effort to find common ground on how best to approach challenging ethical issues (Kahn, Jeffery, Catharyn Liverman, and Margaret McCoy 2014). The committee adopted a principle-based approach in recognition of its accessibility and applicability for its task. The ethics principles and applications are as follows: avoid harm, beneficence, acceptable risk-benefit balance, fidelity, transparency of decision making, commitment to continuous learning, and procedural fairness of decision making (Kahn, Jeffery, Catharyn Liverman, and Margaret McCoy 2014). So while it seems unlikely that we can foresee and plan for all the possible scenarios and issues that the mission will come into contact with, these principals create a guideline for how to proceed. With this in mind, for our mission we want to use the 9 IOM principles stated above to maintain ethical standards during the mission.

Continuing with the idea of ethics and doing what is best for the mission, we were led to a line of questioning about the relationship and knowledge base the mission would have with Earth. We realized knowing too much about Earth may lead to depression or desolation seeing as those aboard will live and die on the mission without getting to experience all that Earth as to offer. However we did not want to lie to those aboard the ship and cause mistrust, so we needed to find a balance seeing as there would be information and data being communicated from the ship to Earth.

With everyone living and working in such close quarters, we realized that we wanted to create a cohesive culture. Chicago Journal writer Dr. Facundo Alonso stated in the Ethics issue that joint action is a phenomenon that is of central importance to our lives here on Earth and that will be no different during the mission (Alonso 2009). There will be a strong reliance on one another aboard this mission and so having a sense of unity and trustworthiness will make the mission stronger. With this in mind, I think it is important to have everyone be on the same page and have the same knowledge in terms of Earth. We do not think only a select few should have the knowledge about Earth nor do we think they should be lied to. While the awareness of Earth and the realization of living solely aboard the mission can be jarring, having mental health workers aboard the ship and also the solace of fellow passengers should be comforting.

Walter Lifton, a scholarly speaker at the American Personnel and Guidance Association Convention, provided concepts that he believes are the basis for deciding appropriate behavior in-group settings (Lifton 1999). He raises many questions about whether you want the group to be task-oriented or people-oriented. We decided we want the people to be both task-oriented and people-oriented because while there is work to be done, the nature of the trip can take a toll on them, making strong relations essential. Lifton also states the importance of having a community work together and that people have the right to dignity within a society that recognizes their differences (Lifton 1999). In the end, Walter Lifton believes that you cannot brainwash for democracy; neither can you correct injustices by depriving others of their rights (Lifton 1999). He brings up an interesting point about people having the right to feel proud about their accomplishments and work and this made me want to bring certain aspects of Earth onto the mission. Lifton's ideas validated our decision to be honest about Earth and took it a step further. We realized we could incorporate Earth-like concepts such as hobbies and activities that make people feel unique in order to help satisfy the longing for normalcy and create a life more similar to one on Earth.

When thinking about the impact hobbies and activities would have on the passenger's state of being, we realized that it could positively influence their work as well. We found that there had been research done to determine if there was a relationship between employee creativity and job-related motivators. A case study of hotel employees in Hong Kong was used and from a sample nearly 1,000 employees, a correlation indicated that there is a relationship between creativity and job-related motivators (Wong, Ladkin 2008). We took this to mean that when employees were motivated, they produced better work. The intrinsic job-related motivators which include opportunity for advancement and development, loyalty to employees, appreciation and praise of work done, feelings of being involved, sympathetic help with personal problems and interesting work, were found to encourage the hotel employees' risk-taking behavior and productivity (Wong, Ladkin 2008). This study helped convince us as to the benefits of allowing for creativity and uniqueness aboard a ship, something that sharply contrasted our initial inclination to promote conformity and have a minimalistic environment.

In the end, we take the health and safety of our passengers seriously and while we cannot predict everything the mission will come into contact with, we can create guidelines to provide an ethical framework for those aboard the mission and also establish a sense of living that is founded on trustworthiness and supportive of individuality. We are hoping that this mission will produce significant findings but are firm in upholding the responsibility to care for those whose lives are dedicated to the mission.

7.3   BIRTH, DEATH AND POPULATION CONCERNS

Along with all the technical problems that come with an interstellar mission, the people aboard are sometimes forgotten about. Resources will be brought based on the population of those who leave earth, but it is important to take into account the fact that throughout history the human population has almost always increased, with higher birth rates and lower death rates, along with longer life spans. The countries of the world are constantly monitoring the population size, and other factors, but in an ever advancing world with boundless resources to find, these increases can almost always be compensated for. These factors are even more important on a crowded ship with limited resources.

This is why population control will be so important on this ship. With too few people the ship can't run efficiently, as well as other more complex issues such as genetic diversity. On the other hand, having too many people will eat up the resources faster than expected, and on a mission such as the one at hand, access to new resources will be next to none. Both options lead to a failed mission which leads to the death of all the people aboard the ship. It's important when considering the options to be aware of the consequences, and that a failed mission in this case will lead to the death of everyone aboard.

This all leads to the best option being a replacement birth rate on the ship. A replacement birthrate means the rate in which a society can maintain the same average population. It would seem like it would be two, being that every two parents needs two kids to replace themselves, but it is actually a little higher do to sickness and other factors.

It seems like it is all figured out, and the problem is solved, but the issue wasn't in the math of how many people need to be born to replace those that die, but in how to make that theory a reality. In the society we live in today, most people get to choose whether they have kids or not and how many they choose to have. In this space ship's society we are talking about a different dynamic. Along with this dynamic, comes the uncertainty of not knowing the outcome.

There is very little data to look at when it comes to controlling childbirth in modern society, but the prime example is in China, with the One-Child Policy. This policy was a set of laws in which China attempted to limit its citizens to one child each to stifle the unsustainable population growth that they were experiencing (Button 2011). In theory, the policy was sound, but the implementation proved very difficult. How could they enforce such a law? In most cases, this led to fines, and in extreme cases, even forced abortions. There were always violations, and the law had many exceptions and loop holes.

The society on the ship would be much smaller than that of China. A thousand people or so would be much easier to manage than the billion people in China, but the fact still exists that some sort of punishment would have to be involved. "In order to enforce such an intrusive, ambitious, and intensive birth plan, the Chinese government has had to become deeply involved in the lives of its citizens, often to the point of committing human rights abuses" (Button 2011). On this ship, there is no doubt that whatever government we have, they will have to be deeply involved in the lives of its citizens, but is this kind of law enforceable without unrest? Unrest is one of the most dangerous things on a spaceship where no one can leave and each person is needed for the whole to survive. The Chinese government had "through enforcement by means of coercion, forced sterilization and abortion, social stigmatization, and other brutal means, China has created a culture of fear and paranoia around pregnancy" (Button 2011). Such a culture would surely end the mission and lead to the death of everyone aboard.

The question becomes then, can this law be enforced on this spacecraft. Much of the answer has to be discussed in the government and protection section of this paper, but there are other ways of possible cooperation. There will only be around a thousand people on the ship, all volunteers at the beginning. Most likely these volunteers will agree to certain laws before electing for the mission. This means that the issue of enforcement may lie with later generations.

It is important to note that for every subsequent generation born on this ship, the culture will change, and society may evolve differently than it does on earth. For example, if the parents all abide by the two children rule, the children may not feel forced, two children will just seem like the cultural norm. This is the best outcome, since if it can be created as a cultural norm enforcement won't be needed, which makes it easier to keep the peace in such a fragile world.

Of course in human history, there has always been the exception: the person who will break the norm, and though it is hopeful that these cases will be few and far between, the laws must be in place to correct them. There will most likely not be any money on this ship, so monetary punishment is out of the question, and is punishment really the goal. The goal is to keep population stable, so maybe corrective action is a better way of looking at it. Once a baby is born, there are few corrective actions that can be taken. It cannot be looked upon as acceptable because then the population is under risk of breaking that societal norm, but killing a child will undoubtedly cause unrest that cannot be afforded.

The key in keeping the population stable lies in this societal culture. Resources and shelter, and the basic essentials to living will most likely be evenly distributed amongst the people, but in the case of a family with more than two children, they should only receive goods for two children. An unfortunate consequence of this will be hardship for these families, but they will be knowingly electing for this hardship when they choose to give birth to a third child. In cases of exceptions such as twins during the second birth, we believe exceptions may be made in the law, but this sort of law will cause very few people to choose a third child. It allows for population control without the use of unnecessary force, since the society itself will control each other.

As stated above the replacement birth rate for a population is actually a little above two children, due to unexpected deaths and other factors. The few who may choose to have a third child can account for this extra piece. The question is then, if the law works as planned, and everyone only has two children, or if people choose to have no children, what is the course of action. Just as the society cannot afford to have too many people, it cannot afford to have too few either. Can we force people to have children? Again, this sort of action will undoubtedly cause the sort of life ending unrest of the ship that is trying to be avoided.

This is where more science and less social science

comes into play. While the societal norms and pressures will hopefully keep any sort of fluctuation in population very small, there must always be a backup plan when there are such high stakes. In this case, when there are too few births the answer lies with Assisted Reproduction Technologies (ART). On earth, this technology is mostly used for those who are sterile and cannot have children of their own. There are multiple types of ART babies, which include artificial insemination (where the egg and sperm are surgically placed inside a woman's uterus), as well as using artificial wombs where no woman is needed after acquiring the egg (Lambert 2003). It is suggested that in emergency situations, the ship have in storage both human eggs and sperm, in order to artificially birth children to keep the population at a stable level.

With so many factors to attempt to control, the question is why can't we rely on science in more than just emergencies? Why attempt to control birth rates at all if we can simply choose how many children to "make"? It's not that simple. There are too often complications when it comes to these processes. While it would make controlling the birth rate much easier, there is no guarantee in a healthy population. "Assisted reproductive technologies babies are at higher risks of cerebral palsy, premature birth, low and very low birth weights and multiple birth defects, specifically cardiovascular and musculoskeletal," all of which are things that would be too much of a task to treat on the spacecraft considering some of the lifelong treatments necessary with certain defects (Lambert 2003). While the same risks are always involved with natural birth, the risks are twice as high when dealing with ART babies (Lambert 2003). The limitations then of this science, are what cause its use to be for emergencies and not everyday child birth.

The hope is that through societal pressures created by the norms, starting with the original crew, most of the population control can be achieved with law enforcement and science there to push it back on its course in the case of fluctuation, but the importance of peace should be stressed. Peaceful coexistence must be the number one priority, and it must be considered when implementing any laws on the ship.

Along with keeping our population at a level that can be provided for, there are other things about the society that have to be taken into account. In this enclosed society in the spacecraft, every job and person will be necessary to keep the system moving smoothly and keep everyone alive. With limited resources, there is little space for people who cannot, and maybe will never again be able to work. As harsh as this may sound, failure of the mission will mean death to everyone aboard.

The question then lies, what level of cannot work can be accepted, and what is done with those who cannot meet the standard? First, it is important to realize the different categories of people that are being considered. The major dependency groups are the young, the old, and the sick or disabled, and they all should be measured and dealt with differently (Illsley 1981). The young will eventually be productive, and are an investment in the mission's future. This leaves us with the sick or disabled and the old.

When looking at those who are sick or disabled, the first question is: is it long-term? Next, it has to be determined to what level they are sick or disabled. Will it impair working, and how much? There are many illnesses and disorders on earth, and any of them could also present themselves on the ship. The ship will have very limited medical supplies that must be conserved when possible. They will have to treat what can be treated easily, but for those things that cannot, the ship must be protected from disease. One viral disease has the potential to kill everyone on the ship, even those which may be treatable on earth.

Each situation has to be evaluated differently. Those illnesses that can be treated quickly, easily, and with little resources will be treated and the people will be sent on their way to continue daily tasks as soon as able. Any highly contagious diseases will have to be quarantined as soon as possible to protect the ship. Lastly, the limited medical staff cannot be wasting precious time and resources on fatal, long-term, or debilitating illnesses. Sadly, the list of fatal and incurable diseases will be far longer on this spacecraft than earth.

Just like minor illnesses will be worsened by the lack of resources on the ship, so will disabilities. Those who become disabled to the point where they can no longer work will be a strain on the rest of the population of the ship. Raymond Illsley, a professor in sociology describes these dependent groups:

> "The handicapped, the chronic sick and the elderly are described by the author as 'dependency groups' to emphasize, despite their different diagnostic labels, their common status as citizens and patients. They share several crucial characteristics—they are not amenable to curative treatment and not being susceptible to professional skills are professionally uninteresting; they are potentially costly as long-term users of medical and social services; having multiple needs, they are not the clear responsibility of any one service; they are economically unproductive and hence economically and socially dependent" (Illsley 1981).

When Illsley talks describes these "dependency groups" as "professionally uninteresting", he is referring to their ability to go back to work in a normal earth society. In this case, he is talking about their dependency on "medical and social services", meaning the government in most cases. On earth, in countries like ours the dependency strains government, but that strain gets divided amongst over 100 million tax paying, working people. On this ship, the strain would be divided amongst 1000 people, while also eating up more than just medical and social services, but food and shelter supplies that will be limited and need for productive workers.

The old are the last group that can be classified as dependent, but everyone ages differently, and how do you put a limit based purely on age? An economic study done on Productivity and Age concluded that "average daily hours of work for both men and women peak at around 8 hours per day for adults 30 to 49 years of age, [and that] average hours worked at jobs are fairly stable after school years are completed and begin to decline at age group 55 to 59," but these are still averages, and some people remain fit into their 80's and longer (Guimarães 2011). The study also claims that women's productivity decreases more slowly over time, but that this has to do with their jobs along with their higher average lifespan (Guimarães 2011). The study even admits to "underestimate the productivity of older adults" since it is rapidly increasing (Guimarães 2011).

It comes down to that there is no age where a person stops being productive, but that illness and disability cause the decline. An article in the scholarly journal "Nature" presents the problem that "as life expectancies around the world continue to climb, whether longer life means, on average, a longer period spent suffering from age-related frailty, disability and disease" (Kirkwood 2008). Disability and disease come with age, and that is what causes productivity to slow. There is no age where all people die because it is not age that hurts someone, just the large susceptibility to illness. Because of this, the same theories for illness and disability can be used for everyone entering old age, with no set lifespan or age limit.

What do we do when a person becomes disabled or

sick to a point where they will never again be productive? The horrible truth is that on this mission, these "dependents" become a liability. On earth, the reason these people survive is due to a code of ethics that every life is worth saving. On the ship, these ethics change. We are not claiming that human life is no longer important, but only that the situation and stakes are different. On earth, it is only one life in the balance, but on the ship every life is endangered every day. Every decision could lead to the death of every person, and every sick person that resources are wasted on puts them all at risk.

In the journal of Medical Ethics, C.S. Campbell presents six questions to be answered before taking a human life, with the most important being "Have all alternatives to obtain this purpose, short of taking a life, been exhausted?" and "What is the purpose?" (Campbell 1992). In the extreme situation that this society will be in, there is a clear purpose, and there are no other options. While this will never make taking a life okay, it seems necessary, and will save the lives of everyone else aboard.

7.4   TRAINING AND OCCUPATION CONCERNS

The first question that will be addressed in this section is what types of jobs are necessary to be self-sustaining. This question is important because work has to be done on the ship for the mission to succeed. Sanitation, technology child-rearing etc. are all things that require people to work. Thus the occupations that we decided were necessary for the mission to succeed and for the society on board to function are as follows; scientists, linguists, researchers, waste management, a cleaning crew, supply management, engineers, maintenance, navigators, pilots, caretakers, doctors, nurses, health care providers, cooks, protection forces, farmers, agriculturalists, and teachers.

There was little direct research on specific jobs

necessary for a society to be self-sustaining. However there were a lot of studies done on communes and enclosed societies. The ones that worked emphasized a strong work ethic, that the needs of the group outweighed the needs of the individual; the communities relied on themselves to do the work needed. "In the kibbutz, however, in which it was axiomatic that members place the interests of the group above their personal interests, work was motivated by hakkarah, or the self-conscious recognition that the welfare of the group depended on each member working as efficiently as possible. The consequence [...] was that virtually all members worked with intense, even obsessive, dedication" (Spiro, 2004). "A second dimension of this value consisted of the norm that if the interests of the group conflicted with those of an individual, the former would take precedence" (Spiro, 2004). This was a Jewish commune being studied. Another group that was studied was the Amish for "they contend that each individual has the obligation to work and be productive. Idleness, waste of time and frivolousness are all strongly condemned. [...] Work should be communal, intertwined with the community, and never a source of individual pride and exhibition" (O'Neil, 1997). Both of these groups were successful in being self-sufficient in their work. And as one can see the overlapping point is the emphasis on working hard for the greater good. Thus it will be necessary for our mission to strongly emphasize this idea as well, that everyone needs to work for the greater good so that the mission will succeed. How this idea will be emphasized will be discussed later.

A different article discussed the need for a balance of intellectuals and physical workers so that both kinds of work can be done. "During the nineteenth and early twentieth centuries, many tolerant and secular communes were attempted. Only a few were even moderately successful. Because they are based on the promise of materialistic advantage, they attracted many followers who disdain the sacrifice necessary to eke out a living in relative self-sufficiency. They also attracted too high a ratio of intellectuals to workers and were troubled by constant infighting." (Thies, 2000) Thus we cannot just have scientists come along for the research aspects of the mission but also workers who will clean and people to take care of the children etc. There needs to be a diversity of occupations on board to meet the needs of the ship. Again this is why we chose the aforementioned list of jobs.

Another aspect to consider with having different occupations on board (especially menial versus intellectual) is that it could lead to class distinctions. To have only scientists would be an obvious folly. We cannot just have intellectuals on board. The problem with having menial workers though is that it could cause class distinctions and tension between those directly working on the mission and those who are just on board. We need jobs to maintain a sanitary environment, functioning machinery, etc. They do actually mean just as much as the scientific jobs. We address the issue of class/occupational distinctions in our plan for the education and political system. However, this part of the essay will continue by focusing on the education system.

The question of how the education system will work is important because as mentioned earlier, it will form the occupational system as well as the overall mentality of the members of the mission. So that is why though these categories of jobs will be assigned with the original crew, later the jobs will be given based on tested capabilities and success through an apprenticeship program. This seemed the most compatible with what we are trying to do; create oligarchy groups out of the jobs. This kind of program would give hands on experience to the people as to how to do the jobs well. This way the students could have very early training, and we could quickly gage whether someone fit in that field or not. Also, indoctrination into the working for the greater good mentality, as well as occupational pride can be developed early.

There are two ways that the apprenticeship program could work. The first is the traditional way that at an appropriate age, a student would begin to learn in a hands-on setting from one master. The problem with this is that it is not effective in its use of time or resources. With one apprentice to one master, more masters would be needed to teach multiple apprentices. This leads to the second option, which is an apprenticeship/classic classroom combination. Basically there is one master who has multiple students, or apprentices. He or she teaches the students all together but still in a hands-on way. This way worked for an apprenticeship in China where there was one master and 20 apprentices, where "over time, they [the apprentices] develop a unique perspective on their topic as well as a body of common knowledge, practices and approaches. They also develop personal relationships and established ways of interacting. They may even develop a common sense of identity. They become a community of practice" (Gowlland, 2012). This combination worked very well for this group and was very efficient in its usage of time and human resources. This quote though also brings up another important point about the community aspect of it.

Apprenticeships create a personal bond that does not always form in a teacher to student relationship. This is because "the relationship of the master craftsperson to the apprentice played a vital role. With care and nurture, in many cases the master craftsperson often took on the role of a parent -- teaching character, morals, ethics, and integrity, all while mentoring the young worker in the traditions of the occupation and, more importantly, adult life. In essence, traditional apprenticeships assisted young workers in transitioning into adulthood" (Christman, 2012). This is crucial when understanding the need to emphasize a strong work ethic and unifying the occupational groups for the oligarchies. The masters have a personal connection with their apprentices because they work with them side by side, not just talking at them like a teacher. The master can teach much more than the skills for the job but also ways of living, ideologies, how to be a good member of the mission, etc.

Another article discusses the history of apprenticeship which shows the same thing. For a long time apprenticeship was the preferred way of teaching especially because most jobs were crafts that required masters of those crafts. It is noted that apprenticeship was not just teaching a skill but also teaching one how to live life, and instilling a set of values into the next generation. The masters work very closely to not only teach a skill but ways of thinking and comportment as well. It discusses how apprenticeships lost favor with the dawn of industrialization. Work became efficient and mechanized no longer requiring a real set of honed skills. In turn education became about facts learned in a classroom (Wang, 2009). So again, an apprenticeship program could allow for greater community pride and adherence to the rules. This kind of program creates a bond amongst the people in it and a sense of pride as opposed to just telling them facts.

Also, as stated before, the members on board will learn their craft better. A study of apprenticeships to scientists showed that "several learning outcomes associated with apprenticeships including increased understandings of science content and nature of science as well as development of interest in science related careers." (Burgin et al., 2012). This article shows that overall the students of science who learned through apprenticeships had an increased understanding of the material they were learning than students who just learned it from a text book. This can be applied to other mediums as well. The members on board will learn better hands on rather than just in a class room setting.

In conclusion for this mission to work it is necessary to have a diversity of occupations to meet the needs of the ship. These occupations can cause divisions so an apprenticeship program will be built to instill ideas of working for the greater good, not one's own edification, as well as pride on whatever work one has been deemed able to do. However a hybrid traditional classroom size of about 20 people to one master will be more efficient. This will all help to form the members of the ship to be capable and ready to work for the mission.

7.5 PHYSICAL/PSYCHOLOGICAL QUALIFICATIONS

The objective of this section is to address only a mere fraction of the sociological concerns a long duration space flight into interstellar space would entail.

First, it is important to consider the practical questions of what logistics are involved in the recruitment process and how one would go about the screening of individuals who are undertaking this mission. This field of questioning naturally lends itself to a widespread number of issues including selecting an official language, deciding on crew requirements and creating a plan to provide facilities to help in the maintenance of physical and mental health of the crew members.

The course of action taken in recruiting the crew is vital to the success of this mission. Given that this mission will be multi-generational, it is imperative that all preemptive action is taken to ensure that the healthiest generation possible is leading the way. While it is acknowledged that diseases cannot be entirely prevented, it can be controlled.

A five hundred passenger mission requires a very organized recruitment and screening process. A mental health evaluation would check for mood disorders including manic depression and schizophrenia. Also, preventative action should be taken wherever possible to ensure that the crew members are not predisposed to severe mental problems including major mood and thought disorders (manic depression or schizophrenia) have not been reported during space missions. This likely is due to the fact that crew members have been screened psychiatrically for constitutional predispositions to these conditions before launch, so the likelihood of these illnesses developing on-orbit is low.

It is also important to identify specific qualifications and skill sets required on board the spacecraft. These jobs would include everything from nuclear physicists to waste management custodians.

The outcome of the mission has a better chance of success if we open the brain pool to be an international effort. This decision raises the question of what the official language will be. It will be important to establish a universal language in order to maintain efficiency and continuity over generations. There are often minor translation differences which can distort the meaning of research. This language will be chosen while on Earth.

The interstellar spaceship mission to Kepler 186f will entail a lot of mental and physical stresses on the crew. It is important to devise a plan on how to best address all ailments which may arise. We must be mindful of providing resources which are both cost-effective and humane.

Even on Earth, sleep is acknowledged to have a profound effect on a human's function in daily activities (Fucci, 2005). All personnel will need to get the optimal amount of rest in order to contribute to the goal of planet exploration and colonization. "Circadian disruption and sleep loss have been documented in astronauts on missions in space as short as 10 and 16 days. The resulting diminishment in alertness, cognitive ability, and psychomotor performance can pose a serious threat to the safety of the crew and the vehicle, as well as the overall success of the mission" (Fucci, 2005)

"Challenges include achieving proper temperature and humidity balance, maintaining a continuous supply of fresh air, ample water, developing appealing and easy to prepare foods that are acceptable in many different cultures, and improved hygienic systems that are easy to use and perform well under conditions of microgravity. Quality sleep is important for professionals to do their jobs properly, and for passengers and settlers to react appropriately to emergencies, enjoy their flights and reach their destinations reasonably refreshed and with their performance capabilities intact. Sleeping quarters may require better shielding from sound, light, and adjacent activities. Maintaining a 24 hour day by means of illumination control and activity scheduling will help. Like cruise lines, space tourism companies that earn a reputation for providing quality experiences will develop a competitive edge over those that offer the minimum."

"Eventually, settlers will need all of the major social institutions that support societies on Earth [17]. These include government, an economy, nuclear families, an educational system, a legal system, and if not organized religion then something that provides a coherent worldview and set of values and caters to spiritual needs. In some cases social institutions may be exported from Earth (for example during a religious migration), in other cases these may emerge in skeletal form (for example, martial law). First and foremost, in all cases institutions will require modification to meet the demands of crew demographics and the local space environment." (Harrison, 2010)

This source outlines recommendations for a three step process in coordinating efforts for a global space exploration mission. This could inform decisions having to do with the idea that we would simulate a model mission on Earth before sending recruits into space.

Next it is important to consider the psychological and physical human limitations in long duration space excursions.

After much consideration, we propose a multi-staged dry run. The money and time saved in the long run would far outweigh the disadvantages of a less hastened mission to Kepler Planet 186f. A dry run would allow social scientists to test the limits of the human psyche (Seguin, 2005).

This action plan would commence with a mission to Mars and then next Alpha Centauri. Finally, after informed improvements have been made, a crew will embark on their journey to Kepler Planet 186f.

Furthermore, it is recommended that religion and traditional forms of entertainment continue to provide an outlet for society even in an extraterrestrial setting. The humans on board the spacecraft will be subject to extraterrestrial stresses and it is important that outlets be provided in order to maintain the mental health of the crew.

It is important to note that our discussion of the sociological implications of an interstellar mission is not sufficient but merely a starting point for further research and preparations. In fact, there is little to no solid knowledge about the physical and psychological impact of long-duration microgravity and the high radiation that occurs in deep space.

The most pressing goal is to commence with a mission to Mars. This discussion could benefit from further analysis on interpersonal interactions and on the fear of competition and isolation a long duration mission.

7.6 REFERENCES

Alonso, Facundo. Ethics, Chicago Journal. 119.3 (April 2009), pp. 444-475.

Ansdell, M. Stepping stones toward global space exploration. Acta astronautica. (06/2011) , 68 (11-12), p. 2098 - 2113.

Burgin, S., Sadler, T., & Koroly, M. (2012). High School Student Participation in Scientific Research Apprenticeships: Variation in and Relationships Among Student Experiences and Outcomes. Research In Science Education, 42(3), 439-467.

Button, Graham. "China's one-child policy and the population explosion." Indian Journal of Economics and Business 10.4 (2011): 467+. Global Issues In Context. Web. 1 Dec. 2014

Campbell, C. S. (1992). "Aid-in-dying" and the taking of human life. Journal of Medical Ethics, 18(3), 128–134.

Cardoso, A., Guimarães, P., & Varejão, J. (2011). Are Older Workers Worthy of Their Pay? An Empirical Investigation of Age-Productivity and Age-Wage Nexuses. De Economist (0013-063X), 159(2), 95-111. doi:10.1007/s10645-011-9163-8

Christman, S. (2012). Preparing for Success through Apprenticeship. Technology & Engineering Teacher, 72(1), 22-28.

Dick, Steven J Interstellar humanity. Futures : the journal of policy, planning and futures studies. (08/2000) , 32 (6), p. 555 - 567.

Fucci, R., Gardner, J., Hanifin, J., Jasser, S., Byrne, B., Gerner, E., Brainard, G. (2005). Toward optimizing lighting as a countermeasure to sleep and circadian disruption in space flight. Acta Astronautica, 1017-1024.

Gilster, P. "The Ethics of Interstellar Journeying" Centauri

Dreams. (2008).

Gowlland, G. (2012). Learning Craft Skills in China: Apprenticeship and Social Capital in an Artisan Community of Practice. Anthropology & Education Quarterly, 43(4), 358-371.

Harrison, Albert A. Humanizing outer space: architecture, habitability, and behavioral health. Acta Astronautica. (03/2010) , 66 (5-6), p. 890 - 896.

Illsley, R. (1981). Problems of dependency groups: The care of the elderly, the handicapped and the chronically ill. Social Science & Medicine. Part A: Medical Psychology & Medical Sociology, 327-332.

Kahn, Jeffrey, Margaret McCoy, Health Standards for Long Duration and Exploration Spaceflight. Institute of Medicine (2014).

Kirkwood, T. (2008). Gerontology: Healthy old age. Nature, 739-740.

Kramer, M., "NASA Mulls Ethics of Sending Astronauts on Long Space Voyage." Space.com. (2014).

Lambert, R. (n.d.). Safety Issues In Assisted Reproductive Technology: Aetiology Of Health Problems In Singleton ART Babies. Human Reproduction, 1987-1991.

Lifton, W. "Can You Brainwash for Democracy?" Rochester, NY. (1999).

O'Neil, D.,J. (1997). Explaining the amish. International Journal of Social Economics, 24(10), 1132-1139.

Seguin, A. (2005). Engaging space: Extraterrestrial architecture and the human psyche. Acta Astronautica, 980-995. Retrieved December 1, 2014, from Science Direct.

Simon Wong, Adele Ladkin, Exploring the Relationship Between Employee Creativity and Job-Related Motivators in the Hong Kong Hotel Industry. International Journal of Hospitality Management

21.3 426-437. (2008)

Spiro, M. E. (2004). Utopia and its discontents: The kibbutz and its historical vicissitudes. American Anthropologist, 106(3), 556-568.

Thies, C. F. (2000). The Success of American Communes. Southern Economic Journal, 67(1), 186-199.

Wang, Victor (2009). Definitive Readings in the History, Philosophy, Practice, and Theories of Career and Technical Education. Zhejiang University: Zhejiang University Press.

Williams, D. Acclimation during space flight: effects on human physiology. CMAJ. Canadian Medical Association Journal. (06/2009), 180 (13), p. 1317 - 1323.

Chapter 8: Communications and Navigation
By Kristen DiMarino and Ryan Vitter

8.1 INTRODUCTION

This section examines communications and navigation issues. Included among the communications issues are the questions of how to send back information to Earth; what information needs to be sent back to Earth; and communications with any other intelligent life. These are important aspects of the mission. Without communications and navigation, the mission cannot get off the ground and cannot successfully reach its destination. The communications and navigation department of a multigenerational starship will control propulsion, interstellar beacons, and contingency plans.

Propulsion alternatives were investigated in Chapter 1 of this volume. Solar sails and nuclear fusion power were among those proposed as viable methods. We concluded that fusion power seemed to be the most likely choice for the mission. With respect to navigation through space, we concluded that a pulsar-based system of navigation would be best. With respect to the type of communications to be utilized, we concluded that the mission should first use radio communications and then as we get farther away from the Earth, the mission may convert to the use of lasers for communications. Regarding the kind of information that should be sent back to Earth, we concluded that this depends on how far away we are from Earth at the time of communication. Obviously, the farther away we are from Earth, the longer it will take for information to be sent back. Actually, there is the question of whether we should communicate with Earth at all? One view was that communicating with Earth during the course of the mission would not be worthwhile. However, once the mission reaches Kepler 186f, we concluded that the participants must send back a beacon to Earth to inform the planet that we arrived. What if any intelligent life is encountered? Should we communicate with it? We likely will not encounter intelligent life, but if we do, we will have to work together with the other intelligent life.

We also considered the possibility that we may encounter microbial life forms. After reviewing other studies on keeping spacecraft sanitary and avoiding contamination of outside particles, we found there are many sanitation techniques to prevent this.

We also examined contingency plans for possible disaster scenarios such as failure of power and being thrown off course. One approach was to have a spare spaceship completely loaded with fuel alongside the main ship as well as spare parts for repair.

Regarding navigation, we considered ways of using pulsars and quasars to maintain the correct course. Finally, we briefly reviewed what skills are necessary for the communication and navigation crew, especially the ship's navigator. Leadership and effective communication skills are absolutely necessary.

8.2 COMMUNICATIONS IN SPACE

Today we strive to acquire the knowledge of the universe. We set out to answer many questions that lie in space, solar systems, black holes, and finally the universe that holds all these compelling and mind-blowing conundrums within. Technology is accelerating forward at a great speed and we see new things being invented and improved. For many people this represents the future in which humans set out to brawl all questions and try to find answers that will help generation ahead. Today, the general public is even thinking about interstellar travel. Of course many things would have to be taken into account in order to make that idea plausible and a reality.

One of the biggest problems in creating a spaceship that would have the ability to go on a multi-generational mission, and have the ability to travel at a high speed, are communications and propulsion engines. In the next few pages we will look at some of the ways we could communicate in space to great distances. We will also take a look at the kind of information that would be useful to send.

Even with today's technology, it would require an immense effort and time to communicate with any spaceship that is light years away. For instance, visible light and radio frequencies travel at the speed of light, this is the maximum speed that anything can travel. So if we were to assume that a spaceship was on its path to another solar system or star and was already a light year away, it would take a whole year for a signal sent from Earth to be transmitted and received on board and vice versa (Satellite Reception, Interstellar Communication p.1, End-to-end Interstellar Communication System Design for Power Efficiency p.18-20). Let's examine some pros and cons of using radio signals.

When using a radio to communicate with the spaceship travelling at least one light year away from Earth, there is interference in space that causes the signal to dissipate, since the distance is so great (Interstellar Communication p.1, 2, Life in the Universe p.409-411). Information size would have to be constrained so that energy consumption is reduced. Nevertheless the chance of the signal and information being lost will still exist. Also, because of such great distances, it will require a lot of power and bigger antennas than already exist, in order to transmit the necessary information. It was discovered that some of these problems might be solved if broader and wider bandwidth are used. Therefore it will also result in a simpler design (Interstellar Communication p. 2, End-to-end Interstellar Communication System Design for Power

Efficiency p.18-20, Life in the Universe p.409-411).

A positive aspect of using radio waves to communicate is that information can be transmitted at a maximum speed from point A to point B and information may be more complex and detailed when comparing to a laser communication (End-to-end Interstellar Communication System Design for Power Efficiency p.18-20, Life in the Universe p.409-411).

Another possibility of communicating in space at great distances would be by using a high-powered laser. It would have the ability to be detected on Earth and the light beam would travel at the speed of light. It has been discovered that it takes less energy output for a laser to outshine a star, therefor making it fairly easy for the light signal to be detected on Earth (Results of Kirari Optical Communication Demonstration Experiments p.35-37).

A number of SETI (Search for Extraterrestrial Intelligence) projects, for the past few years have been trying to find extraterrestrial life outside our solar system by searching for signals that may have been transmitted by other civilizations, primarily in the radio frequencies of the electromagnetic spectrum. Recently they started examining a water hole frequency, because it has low background noise, which means that the chance for interference decreases. Many attempts were made to transmit signals to clusters of star and the most famous was conducted in 1973 when a radio message was sent. It was sent to a cluster of stars in Milky way, nearly 30,000 light years away, however even if the extraterrestrial civilization answers to this message it will take another 30,000 years for the signal to be received on Earth, making it a total of 60,000 years (End-to-end Interstellar Communication System Design for Power Efficiency p. 34-36, Sending and Searching for Interstellar Messages p.614-617, Life in the Universe p.409-411).

It is appropriate to conclude that although radio

waves do have the ability to move more information there is a bigger chance of interference, when compared to laser beam. Laser beams may prove to be the optimal communication system, because it requires less energy and although there is a chance for cosmic interference, the chance of disruption happening is lower. To completely answer this question we should say that the optimal plan of communication may be the following: in the beginning of a trip, while the spaceship is still a few thousand kilometers away from the Earth, communication should be made through radio, because time for a signal to reach Earth and vice versa will be substantially lower than if the spaceship was located a few light years away. As spaceship would start to get further away (a few light years away), a laser beam would be an appropriate way to communicate with the Earth, not because it would take less time (radio waves and light move at the same speed), but because there will be less possible interference with the light beam. Finally, as the ship starts to get even further away, the need for communications becomes less, not because there is no need to communicate, but because it becomes pointless (since it would take years for the signal to be received on Earth). It would be appropriate to send a light signal when the ship gets to its final destination (Interstellar Communication p. 2, End-to-end Interstellar Communication System Design for Power Efficiency p.18-20).

Next we will attempt to answer the second question which asks what kind of information should be sent to the spaceship and vice versa. Some of the useful information that may be sent to the ship is maps and coordinates. This, however, would be useful only in the beginning of the trip, when the spaceship is still relatively close to the planet Earth. Also, the ship can send updates once a week or once a month. That would again depend on the distance it will be located from Earth, because information can travel at a speed of light and since this particular spaceship will continue to get further and further away from the planet Earth, time for the information to reach its final destination, will increase.

Probes also can be used to communicate between the spaceship and Earth. It is understandable that it will take an extended amount of time for the information in those probes to get to the final destination. Nevertheless; probes can carry large amounts of data and information. They can be used to carry items such as food supplies and even fuel to the ship in case of emergency. However probes will only be useful at a close distance to Earth, because it would take enormous amount of energy and time for each probe to reach its final destination. So it is one of the possible ways of communicating, however it is not the best (How Does NASA Communicate with Spacecraft).

When we get to utilizing laser beam communication, it is reasonable to say that no real information, such as maps and data will be sent in this process. Laser beams can only be used to indicate an emergency on the ship, to assure people on Earth that everything is all right or to indicate a landing on the planet (Results of Kirari Optical Communication Demonstration Experiments p.35-37).

Technology is accelerating rapidly and every year people are introduced to newer and more improved technology. It may be that by the time an interstellar trip is undertaken, newer technology will become available that will allow spaceships to travel at greater speeds and communication may improve greatly.

8.3   COMMUNICATIONS ON MISSION

For an interstellar mission to a possibly habitable planet outside our solar system, communication is one of the most important considerations. Communication among the inhabitants of the mission ship will arguably be a non-issue as technology for close-range communication between astronauts is already well developed and attainable (NASA Communicating in Space). The issue with communication arises when the mission astronauts must communicate with life outside of their environment. Two important topics that lie under the umbrella of communication outside the mission craft are addressed in this paper. The first, how we will communicate back to the Earth, was addressed in the first portion of the paper. This issue was one of the most controversial communications issues with which we were faced in the creation of this section. The other important question addressed here is how to communicate with extraterrestrial life, if any is found. This mission's success depends greatly on the work of the people in the department of communications and navigation, but what happens once we arrive at Kepler 186f?

When discussing the role that communications with the Earth will play in the mission, we took into consideration the time it would take to send messages from the ship to the Earth as the ship moves away from the Earth. Because of the increasingly long time that it would take the ship to send and receive messages as it leaves the vicinity of the solar system, we came to the conclusion that the use of resources needed to send and receive messages to and from scientists on Earth should not be wasted. Our logic in denying the mission contact with Earth is not only the time it would take to communicate messages back and forth but also, as the mission progresses and the first couple generations die away, the desire to send mundane updates or research back to a planet that the new generations have never seen would fade. We agreed that in order to make the best possible use of our mission and for the mission goal to remain inhabiting Kepler 186f, there would be no need for communication or transmissions of research back to Earth. Though it may seem controversial to cut ties with the ship once it leaves close range of the Earth, it would ultimately be futile to attempt to maintain a two-way conversation through the duration of the mission. However, it would be pertinent to signal to the scientists back on Earth when the mission reaches its destination in order to alert the people of Earth that we have achieved the goal and that Kepler 186f is habitable so that they can then send more missions.

We explored several different possibilities for transmitting a signal back to Earth effectively and with minimal resource expenditure. The first possible option that was considered was using a solar sail to create a signal that would be visible from Earth. Though this idea would be an easily detectable means by which scientists on Earth would be able to find the mission, the idea has several flaws. The source from which this idea came, "Scouting the Spectrum for Interstellar Travelers," discussed the use of a solar sail beacon in an attempt to signal to other civilizations with interest in astronomy, that our civilization is here. This type of communication is the type of signal we would want to send back to the Earth, however, because we cannot rely on the solar wind of the planet's star to be substantial enough to support this type of beacon, it may not be wise to spend resources creating a beacon that may not work properly. The more viable option, which numerous sources discussed in terms of sending signals from Earth to possible intelligent life elsewhere in the galaxy, is to use a cost-optimized interstellar beacon.

The cost optimized beacon could be built on the ship and beamed directly toward the Earth in short bursts that would reach the scientists back on Earth. The beacon would need to be a short pulse beacon in order to conserve power and allow the newly arrived generation who reached the planet to send the signal over a number of weeks or months to be sure it is received (Benford, et. al). Creating a beacon that uses high-power microwaves will be able to traverse the distance from the planet to Earth and will send out a signal in bursts in order to conserve "overall prime power requirement" as well as preventing the beacon from overheating (Benford, et. al 5). Because when the beacon is installed, the direction of Earth will be known, the beacon can be narrowly directed so that the short -burst pulses will be detected relatively easily by receivers on Earth. Benford and his colleagues have created numerous algorithms to discuss the "optimum tradeoff" that comes with building a galactic-scale beacon, "minimizing the cost of producing a desired power density at a long range [to determine] the maximum range of detectability of a transmitted signal" (Benford, et al 2). Using the algorithms these researchers have developed would allow the financial teams and scientists planning the mission to determine how powerful the beacon should be and how much cost is a factor in creating the beacon. Although it need not be explicitly stated, the scientists on Earth will have determined a range of dates during which the mission will likely reach its destination. From that point, they will begin their surveillance of the location for incoming signals. Ideally, another similar beacon would be built on Earth so that when the signal from Kepler 186f is received, a reply signal can be sent so that the extraterrestrial beacon need not be signaling indefinitely.

The other pertinent issue in discussing communication is in regard to possible intelligent extraterrestrial life. The chances that our mission will encounter another intelligent life form that has gone undetected in so close a region to our solar system is unlikely, but in order to be prepared for anything that we may encounter, we will have to have some sort of messaging system to communicate with other life forms. It was suggested that linguists be brought on the mission in order to better the chances that if intelligent life uses their language to contact us we may be able to better decipher it. Realistically, a specialist in human languages would not likely be any more help in a situation like this than any other human. Because we have the ignorance of being confined in the bodies and minds of our own species, we can never decipher the language of another without much cooperation from both sides. Perhaps in this case a drawn plaque similar to the one attached to the Voyager spacecraft would be the best solution. Because of the restrictions on our abilities to prepare for communication with another species, we are only able to educate the rising generations on the possibilities of extraterrestrial life and the protocols on how to approach it. There should be a Planetary Protection treatise created before the mission launches that details how any other life forms should be approached: with scientific respect, a want for understanding, and peace.

The education about how to handle encountering intelligent life should not give our astronauts hope that they will actually encounter any human-like life. On the other hand, it is very possible that they will encounter microbial life on Kepler 186f. Because of its seemingly Earth-like conditions, from what we can tell, it is possible that microorganisms may have developed on this planet as they did on Earth millions of years ago. If this is the case, it is of utmost importance that the generations who reach the planet take great precautions in protecting themselves and the planet they want to inhabit. On Earth, there is a Policy for Planetary Protection that gives instruction to not only prevent contamination between the planet's environment and any spacecraft, but also to preserve future biological development on the planet (DeVincenzi, et. al 15). It will be very important for our scientists to detect life forms on the planet before disturbing the environment to ensure the stability of the planet as well as the safety of the human astronauts on the mission. There are several ways in which the scientists will be able to detect possible life on Kepler 186f before landing, the most promising of which is the technology used to create NASA's Lunar Orbiter Laser

Altimeter (LOLA). This device uses laser pulse beams to model surfaces of planets (NASA: LOLA). If modified to also detect the elements present in the atmosphere of the planet and determine the habitability for creatures we are currently aware exist, then scientists on board the ship can determine where to search further for signs of life. LOLA would also be valuable in finding a proper area where the ship may land when the time comes since it details the surfaces of planets. Several important research documents have been released studying the ways in which cross-contamination can be avoided and how to properly clean devices that may be sent to search for signs of life elsewhere. To prevent false positives in testing for possible signs of life from samples of Kepler 186f, research performed by scientists like Eigenbrode and her colleagues on what chemicals will allow for a full decontamination of sampling tools will be invaluable to the scientists who reach the planet. Current sterilization methods work to keep forward contamination out of samples taken from other planets, but our mission must also consider what effects contaminants from Kepler 186f may have on them (National Research Council et. al 17-19; Mullen). Our mission will have to follow guidelines for the proper sterilization protocols when physically interacting with any part of the planet before the in-flight scientists determine the safety of the biota on the planet, if any exist. Though unrelated as it may sound, these issues are important to the communication department because encountering and addressing extraterrestrial life, no matter how miniscule, is a directive of communication and must be a joint effort among members of the communication, scientific, and protective departments.

Transmitting data to Earth has been determined to be irrelevant beyond a signal of confirmation upon arrival and the likelihood of getting the opportunity to attempt communication with other intelligent species is unlikely, making the journey a lonely one for our generations of astronauts. However, being prepared for the situations that may arise and having protocol to prevent any unnecessary risks that could be taken will make the journey a much safer and more comfortable one for our astronauts. Their beacon back to Earth will begin a new chapter as Earth will likely send more ships to help populate and research Kepler 186f and other planets like it in our galaxy. The research and development that our mission will be able to achieve will allow even more generations of humans to spread across the stars and use the precedents set by the mission to Kepler 186f as a guide on how to successfully achieve a multi-generational interstellar mission.

8.4   BACKUP AND RELATED PLANNING

Every mission should have at least one contingency plan for the unfortunate event of something going awry, especially if the mission is taking itself far from available resources and aide as a multigenerational voyage would do. Tasked with considering backup plans for faults in navigation and being thrown off course as well as power failure, this paper will discuss the various problems associated with the mission's chosen methods of navigation and propulsion. As a reminder, this mission is most likely going to use nuclear fusion power as its fuel source and pulsars for navigation. Additionally, we will examine the necessary skills for the navigation crew, specifically for the head pilot.

Because creating a predetermined flight path is incredibly difficult, given the distance from the planet this flight is attempting to travel to, a more flexible approach to navigating is necessary. By using probes sent ahead, the ship can map out current obstacles and the clearest path to take using short-range technology that can communicate with the ship. Should it be thrown off course, it is expected that the probes can recalculate a new path for us to follow. Probes serve multiples purposes, including communications feedback, acquisition efficiency, and military security advantage (Freitas, 1980). Military advantage is especially important to consider if this ship is not equipped with a full arsenal capable of handling a threat from a planet with all its advantage. Probes can also be used for bouncing messages back to Earth if enough of them were left at specific intervals on celestial bodies as a kind of "lighthouse" (Gilster, 2013). For the short-range distance out of the current solar system and slightly beyond, it is possible to use a telescope for some of the path. Bernard Lovell, creator of the Jodrell Bank telescope, made such a large investment to open the world up to a "new window on the Universe" (Smith et al., 2014), and his telescope could be used to map out a kind of predetermined flight path for the first part of the journey.

It is necessary to take into account a backup plan to deal with possible deterrents. There are many obstacles in an interstellar space mission, such as asteroids or space debris. Any number of things can throw the ship slightly off course, and in space, a slight error in angle can lead to massive errors in direction. "Interstellar navigation involves the determination of position and velocity of a spacecraft to predict its future motion. This determination is made by taking a series of measurements to make the most accurate estimate of the ship's current position and velocity, taking into account errors or inaccuracies in the measurements themselves" (Calabro, 2011). Pulse time of arrival can vary somewhat from pulse to pulse, so if this ship does rely on pulsar navigation, any small error could cause the course of the ship to be off (Liu et al., 2010). Liu and his associates suggest using x-ray pulsars with the incorporation of a Kalman filter to counteract the variation in pulse time of arrival, which would effectively eliminate the errors caused be a delay in the pulses (Liu et al., 2010).

In a 2012 international workshop, an experiment was performed to confirm this method of correcting time of arrival error with the Kalman filter (Du et al., 2012).

Another issue to consider while thinking of being thrown off course by obstacles is the maneuverability of the spacecraft, especially in regards to its speed. If thrown off course, it becomes incredibly difficult to maneuver a starship going at a relativistic velocity, as "such a sidewise cruising at a relativistic speed would impose a significant technical problem to protect the ship and the engine against hard radiation of the ultra-relativistic flow of hydrogen and helium atoms of interstellar gas (relativistic wind) not mentioning possible dust granules" (Semyonov, 2012). Besides the obvious slowing down to make adjustments and avoiding sidewise cruising, course corrections could also be made with powerful laser thrusts as well (Gilster, 2013), and pulsar navigation could be used for position verification and "GPS-like" tracking so that the correction is accurate (Gilster, 2009 & 2010). Quasars are also an option for correction the aberrational errors in navigation (Calabro, 2011).

When considering a power failure, the ship must have some kind of backup for fuel and propulsion. If the ship uses mainly a solar sail, which it could, since "two of the systems, the laser-propelled lightsail (LPL) and the particle-beam propelled magsail (PBPM), appear to be technically feasible and capable supporting one-way probes to nearby star systems" (Andrews, 1994), then nuclear power fusion could be stored as a backup method on the ship. "The total beam energy requirement for an interstellar mission is roughly 1020 joules, which would require the complete fissioning of one thousand tons of Uranium assuming thirty-five percent power plant efficiency" (Andrews, 1994). These numbers imply that should there be a defect in the sail; the ship would cease to have the resources available to fix itself. Other alternatives include storing extra fuel and replacement generators on the spare ship that will accompany the larger ship. This suggestion would be most useful is nuclear fusion power is selected as the primary power source, as it may be only a replacement part that is needed to fix any problems.

Autonomous pulsar navigation is possible by laying down pulsars on celestial bodies (Gilster, 2013), and there are also millisecond pulsars for even more precise "GPS-like" locating (Gilster, 2009). By using pulsars for navigation, the mission will eliminate human failures in judging courses and adjustments. "Navigation by stars and pulsars will be, apparently, routinely used for interstellar flights. Instead of knowledge of winds and local sea currents important for the sea-farers, the astronauts will need knowledge of the peculiar relative motion of stars in the Galaxy to make the necessary course corrections" (Semyonov, 2012). All this being said, a pilot must be capable of making the calculations necessary for even the most minute course corrections. An autonomous x-ray pulsar-based navigation system will do most of the work, especially given the improvements to current x-ray technology combined with clock error and clock drift error adjustments to eliminate planetary ephemerides (Wang, 2012). Some have suggested making advanced autonomous systems that a pilot would simply be able to look over and watch instead of manually run (Garrett, 2012). Even with completely automated systems, however, a pilot should be responsible for knowing how each system functions and for being able to instigate an override, should said systems fail. The ship will have engineers aboard to perform the more difficult repairs and the pilot does have the ability to delegate to them, but the pilot should be able to perform simple corrections on his own.

Although the head navigator may not be the head of the entire ship, his role is crucial and necessary to the well-being and success of the entire spacecraft. That being said, anyone who is meant to fill this role should be screened for mental stability and capability of handling decision-making. The pilot should also possess excellent leadership and communication skills, since the ability to lead and communicate effectively with the members on the team is absolutely essential. It is also a possibility for the pilot to be trained in some basic linguistic communication, since he would most likely be the first one to see any communication in physical form.

To review, there must be backup plans for both navigational error and power failures. Possibilities include creating a short-distance pre-determined flight path, using probes as guides, corrected pulsar-navigation techniques, and quasars. Although the ship will be most automated, the pilot is still responsible for knowing how everything runs in the event that technology fails. Additionally, the pilot must be a well-balanced person and exhibit leadership and communicative skill, and maybe some linguistic abilities.

8.5   NAVIGATION CONSIDERATIONS

Among scholars there is less debate on how to navigate in interstellar space than there is on propulsion. There is a lot of hope in the use of pulsars to help ships navigate across the vast expanse of space. A pulsar is the collapsed leftovers of a massive star that went supernova and became a neutron star. Pulsars have strong magnetic fields which has two directional beams, as the neutron star rotates at a high speed the beams rotate, if a beam points in the direction of the Earth a pulse of radiation can be detected at extremely regular intervals (Astrobio 2012). The use of these pulsars is similar to how GPS works on Earth. There are 24 satellites each contains an atomic clock which uses electronic transition frequency of the electromagnetic spectrum of atoms for a frequency standard (Ray et all 2006). The satellites will broadcast a signal containing the time and a GPS receiver receives the signal from multiple satellites (at least four), and computes the range between it and each satellite based on transmission times and speed of the signal, from these calculations it can determine location (Ray et all 2006). The pulsars would act as natural satellites and would allow for space travelers to calculate the location of their spaceship.

Pulsars have many benefits including its universal availability, zero maintenance, and its ability to be used as an extremely accurate clock. Since pulsars are formed from dying stars there many of them through the galaxy and since they are naturally occurring there is no maintenance needed to maintain them like there is with satellites. Normal pulsars work for about ten million years, there are also millisecond pulsars which are formed when a pulsar has a binary companion that is a normal star; these can last tens of billions of years (Ray, et al., 2006). Furthermore these pulsars make for very accurate clocks,' millisecond pulsars are even more accurate due to increased stability provided by the accompanying normal star that transfers energy to the neutron star which results in a reduced magnetic field which causes the pulsar to spin slower allowing for only a picosecond of change a year in accuracy (Ray et all 2006).

There are also negatives with using pulsars that include less flexibility and the fact they are not as high performing as GPS which allows for less control. Furthermore, it could be difficult to observe pulsars through radio signals in space due to the unpredictability of the interstellar space medium, (Ray, et al. 2006). This would result in a ship needing large antennas to receiver the signals with better accuracy. While this is a possibility it would result in more weight on the ship, maintenance needed on the antennas, and increased vulnerability to the antennas due to the size. Authors Ray, Wood, and Philips offer a promising alternative, observing pulsars through x-rays. Some pulsars are known to emit x-rays which could be detected by much smaller equipment; x-rays are also not affected by the interstellar medium so it would be more reliable (Ray, et al., 2006). However, the authors do admit that this alternative also has its own problems including a much smaller number of pulsars which emit x-rays and a need for an x-ray detector of an unspecified size for extremely precise measurements. Pulsars do seem to be the best option for space navigation, but it is always smart to have backups in case of an emergency.

John Christian in his article, A Survey of LIDAR Technology and Its Use in Spacecraft Relative Navigation, he discusses the use of light detection and ranging (LIDAR) sensors for space navigation. LIDAR is, "an active remote sensing device that is typically used in space applications to obtain the range to one or more points on a target spacecraft", this technology "uses light to illuminate the target an measure the time it takes for the emitted signal to return to the sensor" (Christian 2013). This form of navigation would run into difficulties due to the vast distance between objects in space, but would be a good backup plan.

An even more extreme backup plan would be the use of dead reckoning proposed by John Hemry in his article "Interstellar Navigation Or Getting Where You Want To Go and Back Again (In One Piece)." Dead reckoning uses simple physics to estimate location; this of course would be more prone to human error and should unexpected variables occur it could throw the ship of track (Hemry 2000). There is also basic piloting used by people to navigate before GPS where bearings would be taken off of objects. This is also a possible emergency plan for navigation in space but would be prone to error especially for the fact that most objects in space move making it more difficult to take accurate bearings. These methods of piloting and dead reckoning while useful in older times for navigation should only be used in emergency situations like possible malfunctions of the ships technology.

For this multigenerational expedition to Kepler 186f the choice of propulsion will be nuclear fusion power. With the announcement at Lockheed Martin on their development of fusion within the next decade makes nuclear fusion the most promising form of propulsion that is within the grasp of humans. Should alternative forms like the use of antimatter become a more probable source of energy this could be the energy of the future, but so far nuclear fusion seems like the best option. As far as navigation goes the use of pulsars will be the primary form of navigation with the use of LIDAR and more basic forms of piloting as backups if need be.

Technology is always advancing and more promising forms of and navigation could present itself in the future. All other methods should be examined.

8.6 REFERENCES

Andrews, Dana G. "Cost Considerations for Interstellar Missions." Acta Astronautica 34 (1994): 357-65. ScienceDirect. Web. 29 Nov. 2014.

Astrobio. "A New Way to Keep Clean - Astrobiology Magazine." Astrobiology Magazine. N.p., 9 July 2009. Web. 01 Dec. 2014.

Astrobio. "Interstellar Beacons to Guide Astronauts Across the Universe." Astrobio. 30 Mar, 2012. Web. 14 Oct. 2014.

Benford, Gregory, James Benford, and Dominic Benford. "Searching for Cost-Optimized Interstellar Beacons." Astrobiology 10.5 (2010): 491-98. Web.

Benford, James, Gregory Benford, and Dominic Benford. "Messaging with Cost-Optimized Interstellar Beacons." Academia.edu. N.p., n.d. Web. 01 Dec. 2014.

Calabro, E. "Relativistic Aberrational Interstellar
      Navigation." Acta Astronautica 69.7-8 (2011): 360-
      64. Web of Knowledge. Web. 29 Nov. 2014.

Canganella, Francesco, Petra Rettberg, G. Bianconi, E.
      DiMattia, A. R. Taddei, V. Iylin, N. Novikova, R.
      Fani, P. Brigidi, B. Vitali, C. Lobascio, A. Saverino,
      F. Fossati, and M. Ferraris. "Experimental
      Microbiological Issues Related to Biocontamination
      and Human Life Support inside Manned Space
      Modules." SAO/NASA ADS (2010): n. pag. 38th
      COSPAR Scientific Assembly. Web.

Christian, John and Scott Cryan. "A Survey of LIDAR
      Technology and Its Use in Spacecraft Relative
      Navigation." NASA Technical Reports Server
      (NTRS). 19 Aug. 2013. Web. 14 Oct. 2014.

Crane, Louis and Shawn Westmoreland. "Are Black Hole
      Starships Possible?" arXiv. 12 Aug. 2009. Web. 6
      Oct. 2014.

Devincenzi, D.l., P.d. Stabekis, and J.b. Barengoltz. "A
      Proposed New Policy for Planetary Protection."
      Advances in Space Research 3.8 (1983): 13-21.
      Web.

Du, Jian, Bao-Jun Fei, Ying Liu, and Yu Xiao.
      "Application of STEKF in X-ray Pulsar Based
      Autonomous Navigation." Procedia Engineering 29
      (2012): 4369-373. EngineeringVillage. Web. 29
      Nov. 2014.

Dunbar, Brian. "Communicating in Space." NASA. NASA,
      30 Aug. 2010. Web. 01 Dec. 2014.

Eigenbrode, Jennifer, Liane G. Benning, Jake Maule, Norm
      Wainwright, Andrew Steele, and Hans E.f.
      Amundsen. "A Field-Based Cleaning Protocol for
      Sampling Devices Used in Life-Detection Studies."
      Astrobiology 9.5 (2009): 455-65. Web.

Freitas, Robert A., Jr. "Interstellar Probes." Journal of the
      British Interplanetary Society 33 (1980): 95-100.

Web. 29 Nov. 2014.

Garcia-Escartin, Juan Carlos, and Pedro Chamorro-Posada.
"Scouting the Spectrum for Interstellar Travellers."
Acta Astronautica 85 (2013): 12-18. Web.

Garrett, Henry. ""There and Back Again" A Layman's
Guide to UltraReliability for Interstellar Missions."
(2012): 1-22. Keck: Institute for Space Studies.
CalTech, 30 July 2012. Web. 29 Nov. 2014.

Gifford, Sheyna E. "Alien Atmospheres - Methane, CFCs
and Signs of Extraterrestrial." Astrobiology
Magazine. N.p., 25 July 2014. Web. 01 Dec. 2014.

Gilster, Paul. "Laser Technologies for Starflight." Centauri
Dreams RSS. Centauri-dreams, 22 October 2013.
Web. 29 Nov. 2014.

Gilster, Paul. "Millisecond Pulsars for Starship
Navigation." Centauri Dreams RSS. Centauri-
dreams, 1 June 2009. Web. 29 Nov. 2014.

Gilster, Paul. "Pulsar Navigation: Beacons in the
Darkness." Centauri Dreams RSS. Centauri-dreams,
17 January 2013. Web. 29 Nov. 2014.

Gilster, Paul. "Pulsar Navigation for Deep Space." Centauri
Dreams RSS. Centauri-dreams, 29 November 2010.
Web. 29 Nov. 2014.

Gilster, Paul. "Testing Out Pulsar Navigation." Centauri
Dreams RSS. Centauri-dreams, 12 June 2013. Web.
29 Nov. 2014.

Hemry, John. "Interstellar Navigation Or Getting Where
You Want To Go and Back Again (In One Piece)."
Analof Science Fiction & Fact. Vol. 120(11). 1
Nov. 2000. Web. 14 Oct. 2014.

Herath, Anuradha K. "Migrating Microbes - Astrobiology
Magazine." Astrobiology Magazine. N.p., 15 Oct.
2009. Web. 01 Dec. 2014.

"Homing Signals - Astrobiology Magazine." Astrobiology
Magazine. N.p., n.d. Web. 01 Dec. 2014.

"How Does NASA Communicate with Spacecraft?" How

Does NASA Communicate with Spacecraft? N.p., n.d. Web. 01 Dec. 2014.

Howell, Elizabeth. "Ikaros: First Successful Solar Sail." Space. 7 Mar. 2014. Web. 17 Nov. 2014.

Johnson, Les and Matloff Gregory. "The Interstellar Conspiracy." NASA Technical ReportsServer (NTRS). 1 Jan. 2005. Web. 6 Oct. 2014.

Lesh, J. R., C. J. Ruggier, and R. J. Cesarone. "Space Communications Technologies for Interstellar Missions." Jet Propulsion Laboratory (n.d.): n. pag. Web.

Liu, Jing, Jie Ma, Jin-Wen Tian, Zhi-Wei Kang, and Paul White. "X-ray Pulsar Navigation Method for Spacecraft with Pulsar Direction Error."Advances in Space Research 46.11 (2010): 1409-417. ScienceDirect. Web. 29 Nov. 2014.

Malroy, Eric. "Feasibility Study of Interstellar Missions Using Laser Sail Probes Ranging in Size from the Nano to the Macro." NASA Technical Reports Server (NTRS). 16 Aug. 2010. Web. 14 Oct. 2014.

McGrath, Matt. "'Skunk power' creates confusion over nuclear fusion." BBC News Science & Environment. 16 Nov. 2014. Web. 17 Nov. 2014.

Messerschmitt, David G. "Design for Minimum Energy in Starship and Interstellar Communication." Interstellar Communication (2014): Web. Dec. 2014.

Messershcmitt, David G. "End-to-end Interstellar Communication System Design for Power Efficiency." End-to-end Interstellar Communication System Design for Power Efficiency (2013): Web. Dec. 2014.

Mullen, Leslie. "Alien Infection - Astrobiology Magazine." Astrobiology Magazine. N.p., 25 Aug. 2003. Web. 01 Dec. 2014.

"National Aeronautics and Space Administration." L O L

A. NASA, n.d. Web. 01 Dec. 2014.

National Aeronautics and Space Administration. "The Planetary Quarantine Program." Scientific and Technical Information Office (1974): n. pag. Web.

Obousy, Richard. "Using Fusion to Propel an Interstellar Probe." Discovery News. 5 Apr. 2011. Web. 3 Nov. 2014.

Ray, P. S., K. S. Wood, and B. F. Philips. "Spacecraft Navigation Using X-ray Pulsars." Journal of Guidance, Control, and Dynamics. Vol. 29(1). 1 Jan. 2006. Web. 14 Oct. 2014.

"Satellite Reception." Satellite Reception. N.p., n.d. Web. 01 Dec. 2014.

Semyonov, Oleg G. "Kinematics of Maneuverable Relativistic Starship." Acta Astronautica 75 (2012): 85-94. Web of Knowldge. Web. 29 Nov. 2014.

Shostak, Seth. "The Search for Extraterrestrial Intelligence." Life in the Universe. By Jeffrey O. Bennett. 3rd ed. N.p.: n.p., n.d. N. pag. Print.

Smith, Francis Graham, Rodney Davies, and Andrew Lyne. "Bernard Lovell (1913-2012)." Nature 488.7413 (2012): 592.ProQuest. Web. 29 Nov. 2014.

Task Group on the Forward Contamination of Europa, Space Studies Board, National Research Council. Preventing the Forward Contamination of Europa. Washington, D.C.: National Academy, 2000. Print.

Toyoshima, Morio, Hideki Takenaka, Yozo Shoji, Yoshihisa Takayama, Yoshisada Koyama, and Hiroo Kunimori. "Results of Kirari Optical Communication Demonstration Experiments with NICT Optical Ground Station (KODEN) Aiming for Future Classical and Quantum Communications in Space." Acta Astronautica 74 (2012): 35-37. Web. 1 Dec. 2014.

Vincenzi, D.l. De. "Planetary Protection Issues and the Future Exploration of Mars." Advances in Space

Research 12.4 (1992): 121-28. Web.

Wang, Yi Di, Jun Feng Sun, and Wei Zheng. "Interstellar Autonomous Navigation System using X-Ray Pulsar and Stellar Angle Measurement." Applied Mechanics and Materials 249-250 (2012): 231. ProQuest. Web. 29 Nov. 2014.

Werka, R. O., and T. K. Percy. "Opening the Solar System: An Advanced Nuclear Spacecraft for Human Exploration." Nasa technical Reports Server (NTRS). 24 Feb. 2014. Web. 6 Oct. 2014.

Zaitsev, Alexander L. "Sending and Searching for Interstellar Messages." Acta Astronautica 63.5-6 (2008): 614-17. Web. 1 Dec. 2014.

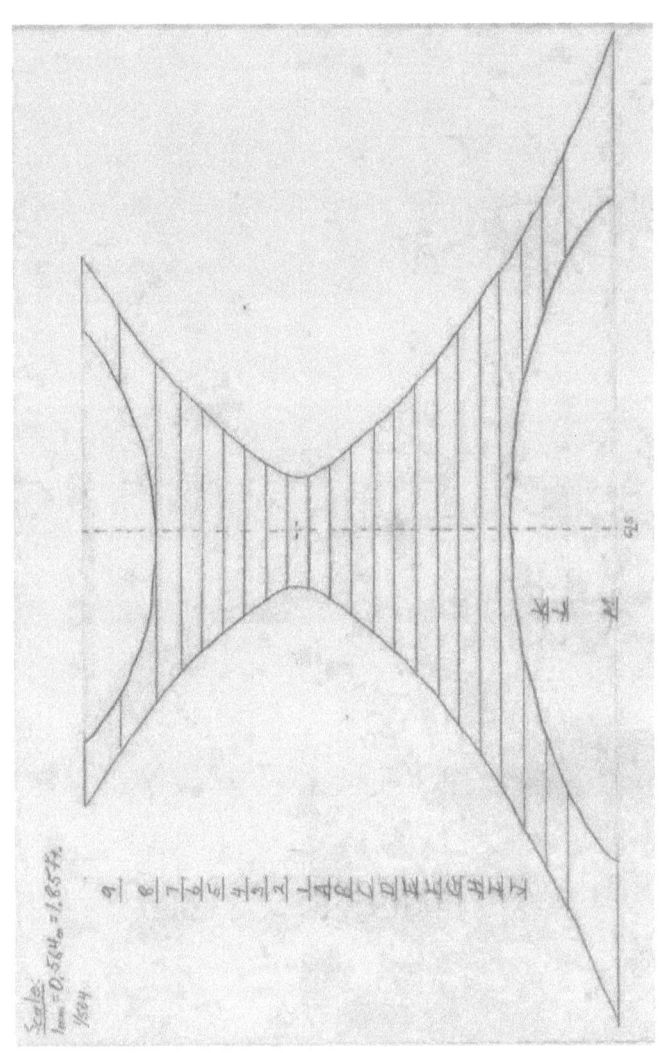

Chapter 9: The Interstellar Medium
By Harold A. Geller and John C. Evans

Any voyage to the stars requires that a vessel and its crew spend most of the voyage in the emptiness of space between the stars. In reality space is not exactly empty. We provide this chapter to give the reader a better understanding of that which exists between the stars.

Interstellar matter is the material lying between stars. Although stars interact with the interstellar medium over the course of their entire lives, we want to divide that interaction into three periods. The first is when stars are born out of interstellar matter which we will discuss in more detail in the next chapter. The second is over that major period of their life when they lose varying amounts of mass to the interstellar medium. Our Sun is undergoing such a loss in the form of the solar wind, which we believe is a process that is not unique to the Sun. We shall inquire into the stellar wind in the first section of this chapter. Finally, the last period is when stars die and is so doing some of them will expel a great portion of their mass out into interstellar space in what astronomers call a supernova outburst.

If you exhale your breath once and let it expand into an evacuated cubical enclosure 1 kilometer on a side, the resulting density of your breath will vastly exceed the density in most parts of the interstellar medium. Although this suggests that interstellar space is nearly a vacuum, there is a significant amount of matter lying between the stars because of the vast volume of space. Interstellar matter is primarily a gas, in which hydrogen is the chief component.

In regions near very luminous, hot stars the gas is ionized, whereas in other regions it is so cold that molecules exist in it. Thus the interstellar medium is far from uniform in its properties. Mixed with the interstellar gas is a very fine dust, whose grains are about the size of one micrometer. Interstellar dust, however, has a very different chemical composition and origin than particles of household dust.

Interstellar matter is not uniformly spread throughout the Galaxy but is clumped together in interstellar clouds that vary in size and the complexity of their association in our Galaxy. The stars of our Galaxy, and presumably the stars in all the billions of galaxies in the Universe, are born in interstellar clouds. And when they come to the end of their lives, many stars throw off matter that mixes with the interstellar medium, where it forms new interstellar clouds and finally becomes the matter composing new generations of stars. In summary stars form from interstellar matter, and throughout their lives stars in turn structure and transform the interstellar medium.

9.1. MASS LOSS BY THE STARS

There are many stars that, for one reason or another, lose mass at one or several times during their lives, or possibly even on a continuing basis. Such mass loss is important for its effect on a star's subsequent evolution, which depends on mass. The solar wind is an outgrowth of physical processes in the corona, and it represents a very small amount of mass loss for the Sun (about $10^{-13}M_S/y$). The evidence that stars are losing matter comes directly from telescopic observations and indirectly from studies of the spectra of stars. For some stars, matter is ejected in one gigantic explosion, such as in supernovae and to a lesser extent novae. Planetary nebulae are another example of a single, but smaller, expulsion of matter by stars. There are also stars for which there is an almost continuous loss of matter, called a stellar wind, over a substantial period of their lives. Stellar winds are probably coupled in part to a range of surface activity, as is the case for the solar wind.

Thus stellar surface activity is the key to understanding a continuous mass loss through stellar winds.

The solar corona is a tenuous, spherical halo of very hot gas whose temperature is 1 to 2 million K. These very high temperatures result from the deposition of thermal energy in coronal gases through the dissipation of energy stored in coronal magnetic fields. This high coronal temperature drives a rapid flow of plasma, composed mostly of protons and electrons, which moves out from the base of the corona as the solar wind. The pressure of the gases in the corona must exceed that of the interstellar medium surrounding the Sun, so that the solar corona continuously expands, being replaced by material from the photosphere. Such a wind is said to be thermally driven.

Several radii outside the Sun, the velocity of the solar wind becomes supersonic and continues to increase at least up to 1AU. By then the solar wind is moving at over 400km/s, eight times the speed of sound in the gas. Because of the Sun's rotation, magnetic field lines, which confine solar wind particles, spiral outward like water from a rotating sprinkler. Perhaps 600,000 tons of plasma leave the Sun every second, which amounts to about 10^{-13} of the Sun's mass per year.

The corona is also subject to sudden and very dramatic disruptions of coronal structure, known as coronal transients. Given sufficient cause, such as a solar flare or other type of eruption, what looks like a magnetic bubble or a series of magnetic loops moves outward through the corona with explosive force. These transients are ejecting coronal matter at velocities of thousands of kilometers per second. The amount of matter being ejected has been estimated to be as much as 10^{16}g, an awesome amount that makes for large variations in the solar wind.

The solar wind rushing out through the Solar System, carrying with it pieces of the Sun's magnetic field, defines a region about the Sun called the heliosphere. The boundary of the heliosphere occurs where a balance is achieved between the pressure of the solar wind and its magnetic field and the pressure of the interstellar medium and the Galaxy's magnetic field. This outer boundary, which at one time was thought to extend only out to about the orbit of Jupiter or Saturn, apparently lies somewhere beyond the orbits of the planets, as more evidence from active satellites in the process of exiting the Solar System seems to indicate.

Do other stars have thermally driven stellar winds? There is no reason to believe that the Sun is unique in this respect. The amount of mass carried away by the solar wind is so small that it would take literally trillions of years for the Sun to lose a significant fraction of its total mass. It is therefore unlikely that we can detect such a low rate of mass loss in other late-type, low-mass stars like the Sun.

Red giants and supergiants are stars with very large radii and very low mean densities. Atomic and molecular particles in their atmospheres are not held so tightly by the star's gravity as such particles are in main-sequence stars. More substantial stellar winds can and do occur in red giants and supergiants, but these stellar winds have smaller velocities than the solar wind. Astronomers have actually photographed a faint halo of scattered light from an enormous envelope of gas surrounding the red supergiant Betelgeuse. This halo is the result of a stellar wind. In fact, Betelgeuse is losing mass at a rate of about 1 solar mass every million years. Astronomers can make out 10-20 micrometer spectroscopic features as well as their changes; detecting episodic pulses of these absorption features.

The recognition that some hot stars of spectral classes O and B continuously lose mass dates from about 1929. Satellite surveys in the ultraviolet part of the spectrum indicate that all stars brighter than bolometric magnitude -6 are losing appreciable amounts of mass through stellar winds. Estimates are that O stars contribute as much as

30% of the matter being lost by stars, even though as a class there are only about 10,000 of them in the Galaxy. Their winds may be driven as much by the pressure of the immense numbers of photons they emit, known as radiation pressure, as by thermal effects.

9.2. EXPLOSIVE VARIABLE STARS

In contrast to a continuous mass loss in the form of stellar winds is the ejection at one time of a star's surface layers to form a planetary nebula. The misnomer was given by the eighteenth-century astronomer William Herschel, who noted the resemblance to the disk of a planet; the planetary nebulae certainly have nothing to do with the planets of our Solar System. In photographs of the Ring Nebula in Lyra (M57) or of the Owl Nebula in Ursa Major (M97) one can see a small, hot, subluminous central star surrounded by a nebulous shell of ionized gas.

The shell is expanding slowly outward at speeds of about 30km/s. The spectrum of its light is an emission spectrum produced by rarefied common gases such as hydrogen, helium, oxygen, neon, and sulfur. The source of energy causing emission from the shell is ultraviolet radiation from the hot central star, whose surface temperature is around 100,000K. At such temperatures, most of a central star's luminosity is composed of ultraviolet photons, and the luminosity is typically 1000 times that of the Sun, but the star's radius is only a few tenths that of the Sun. Such conditions provide the nebulous shell surrounding the star with sufficient energy to give the gas a kinetic temperature on the order of 10,000K, a density of several thousand particles per cubic centimeter, a diameter of several tenths of a light year, for a mass that is a few tenths of a solar mass. The degree of ionization in the surrounding nebula is somewhat higher than that in an HII region, but otherwise they are very similar. Infrared observations indicate that a great deal of dust accompanies the gases in a planetary nebula's shell. Heated by the absorption of ultraviolet and visible photons, the dust in turn radiates in the infrared.

Planetary nebulae's precursors appear to be red giants of moderate-to-low mass located in the Galactic disk. Many stars may have or will eventually become planetary nebulae, although we have identified only about 1000 in our Galaxy. This phase in stellar evolution is brief, lasting a few tens of thousands of years, reducing our chance of seeing it. We can estimate how long a phase in the life of a star is by the relative number of stars that are found in that phase compared to the whole population of stars. An analogy is the problem of how many people are asleep worldwide at any moment. The answer is 1/4 to 1/3 are asleep since the human being is known to sleep from 6 to 8 hours out of each 24 hour period. Turning the argument around, however, if we found 1/4 to 1/3 of the worldwide population asleep at any one moment, then we could conclude that the typical human being sleeps 6 to 8 hours per day. From such reasoning, astronomers estimate that there are actually between 20,000 and 50,000 planetary nebulae in our Galaxy and that a few new ones form each year.

The occurrence of a nova is announced by a rapid rise in a star's brightness, amounting to tens of thousands of times its original brightness in a few hours. This is followed by a slow decline that may persist for a year before the star settles down to its former obscurity. The nova's spectrum shows that matter has been expelled from the star because there is large Doppler shift of the absorption lines toward the blue, indicating a velocity of approach. Soon after the outburst very broad emission lines appear in the spectrum, which indicates that a gaseous shell has been ejected at a high speed, usually a few hundred to several thousand kilometers per second. The broad emission lines result from combining radiation from the front part of the shell, which has a large blueshift, with radiation from the back part, which has a large redshift. The expanding shell has even been seen years later in the case of a few novae. The amount of matter expelled lies somewhere between a few hundred-thousandths and a few tenths of a solar mass. For several novae, radio radiation has been detected, which arises from the thermal energy in their expanding shells of ionized gas. And with the X-ray satellites several novae have also been found to be emitting X-rays. About 30 novae occur in our Galaxy each year; a few even go through recurrent outbursts.

A supernova is an explosion of a star of such immense proportions that it can be observed in an external galaxy even when the rest of the galaxy cannot be seen. These exploding stars suddenly attain luminosities up to several billion times that of the Sun. As many as five supernova outbursts may occur in our Galaxy each century, according to present estimates; most supernovae in our Galaxy probably escape detection because of heavy obscuration by interstellar dust in the Galactic plane. In other galaxies their occurrence varies from several times a century in the brightest and largest spiral galaxies to one every few centuries in the faintest spirals.

A supernova remnant is the expanding shell of gas resulting from the stellar explosion, such as the Loop Nebula in Cygnus. Of the 100 or so supernova remnants that radio astronomers have found in our Galaxy, at least 8 are known X-ray objects and 13 have also been identified optically. Identifying old supernova remnants optically is very difficult because the expanding gas shell thins out so rapidly that eventually it blends with the interstellar medium. Therefore, ages for the supernova remnants that are observed in our Galaxy are probably less than 100,000 years.

The Crab Nebula is a supernova remnant with X-ray

emissions. Part of the X-ray emission is synchrotron radiation from high-energy electrons spiraling around magnetic field lines. The other part is the result of the expanding nebulae plowing into interstellar clouds. Somewhat like the sonic boom of jet airplanes, the ejected shell creates a shock front that compresses and pushes interstellar matter ahead of it. This process heats the intermingling gases to temperatures in the millions of degrees, causing them in turn to emit X-rays.

There are many types of supernovae; we will just focus on the two major types, Type Ia and Type II. The major difference between them is in their spectra and maximum luminosity, but the general behavior of both is pretty much the same. Type Ia supernovae have been observed in all types of galaxies, but they occur most often in the disks of spiral galaxies. Their maximum luminosity is about 4 billion times that of the Sun. For Type Ia supernovae, there is a rapid decline in brightness after maximum luminosity, which is followed by a slowing of the decline with time.

Type II supernovae reach a maximum luminosity of up to 600 million times that of the Sun and exhibit a greater variety of light-curve shapes and spectral changes than do Type Ia supernovae. They appear most often in the arms of spiral galaxies but rarely in elliptical galaxies.

Although both types of supernovae have very complex and variable spectra that are not yet fully understood, they both show spectroscopic evidence for very high expansion velocities, which are on the order of several tens of thousands of kilometers per second.

How much matter is blown off to return to the interstellar medium? The amounts of mass ejected are not known for certain, but the best estimates suggest that Type Ia supernovae are among the older population of stars in our Galaxy and lose about a solar mass of material, whereas Type II supernovae are stars belonging to the younger population of stars and may eject more than 5 solar masses (in some cases 50 solar masses) of stellar matter.

9.3   INTERSTELLAR MATTER

Early in the 20th century astronomers thought that, in our Galaxy, interstellar space was fairly transparent and any dimming of starlight in general could be ignored. Then, in 1930, Robert Julius Trumpler (1886-1956) discovered that open clusters contained fewer faint stars and redder stars the farther away the cluster is from us. For that to be a real effect implied something was very strange about our location. The obvious answer is that there must be interstellar matter lying throughout the region between the stars in the plane of our Galaxy. This matter absorbs and scatters starlight thereby diminishing in brightness and reddening in color (as blue light is scattered more and red light passes cleanly) distant stars compared with nearer ones, which would account for the observations of distant open clusters.

The interstellar matter of our Galaxy, and presumably other galaxies, is a mixture of atomic and molecular gases, mostly hydrogen, along with small solid particles, called grains or dust, concentrated primarily in the plane of the Galaxy. Let us begin by discussing the gaseous component since it is the most abundant part of the interstellar medium.

The gaseous component of the interstellar medium is confined almost entirely to a thin disk in the plane of the Galaxy. In the vicinity of the Sun the disk is only about 1000ly in thickness. About 90% of it, by number, is hydrogen, of which perhaps half is in molecular form and half in atomic form. Atomic hydrogen occurs in both neutral and ionized forms. The volume of molecular hydrogen is low compared to the volume of space with ionized and neutral hydrogen.

The interstellar medium is categorized by the dominant form of hydrogen for the region. As hydrogen is the main ingredient in the interstellar gas, astronomers generally designate a region in which hydrogen is predominantly ionized as an HII region and a region where hydrogen is predominantly neutral atoms as an HI region.

Although most of the mass of interstellar gas is found in interstellar clouds, most of the volume of the interstellar medium consists of warm or hot diffuse gas. Starlight passing through this warm diffuse interstellar gas is selectively absorbed, producing a few absorption lines superimposed on the normal spectra of stars. These interstellar lines can be differentiated from the spectral lines of the O and B stars because they are usually narrow; they are not characteristic of a hot star's photosphere, and they have different Doppler shifts from stellar absorption lines. Interstellar lines are more difficult to identify in stars of later spectral classes, which have many absorption lines. Frequently we see several sets of Doppler-shifted interstellar lines; this means the starlight has passed through several intervening clouds of diffuse gas moving at different speeds along our line of sight.

In the visible part of the electromagnetic spectrum, astronomers have identified absorption lines belonging to such elements as sodium, calcium, and iron and such molecules as cyanogen and methylidine. In ultraviolet spectra, absorption lines for molecular hydrogen, atomic hydrogen, carbon, nitrogen, oxygen, iron, and other elements have been found in data obtained by spacecraft. A surprising find is a large amount of deuterium, the heavy isotope of hydrogen, compared with its low abundance on Earth. Since 1964, in the radio spectrum, discrete lines of hydrogen, helium, and carbon have been observed, lines resulting from electron transitions between energy levels near the ionization limit. For example, a free electron may be captured into level n=110 of a hydrogen atom, from which it can cascade down to level n=109 and emit a photon with a wavelength of 6cm.

Throughout that part of the interstellar medium that has been studied, the abundances of the chemical elements that are heavier than helium are similar generally to what they are in the Sun and other stars. However, since the chemical elements are not spread uniformly throughout the interstellar medium, it is difficult to decide what typical element abundances are. It may be that there are no really typical values. In an HII region nearly 90% of the gas is hydrogen. Another way to compare the relative composition is to say that, for every 10,000 hydrogen atoms, there are approximately 1200 helium atoms, 1 or 2 carbon atoms, 1 or 2 nitrogen atoms, 3 or 4 oxygen atoms, 1 neon atom, 1 sulfur atom, and lesser numbers of atoms of heavier elements, particularly iron.

Another means of exploring the interstellar medium and its structure became available to astronomers in 1951, when a spectral line at 21cm (1420Mhz) produced by neutral hydrogen was discovered. How are photons formed that have a 21cm wavelength, which is in the radio region of the electromagnetic spectrum?

As an electron revolves about a proton, it and the proton also spin like tiny rotating tops. Once in 11 million years, on average, an electron, if spinning in the same sense as the proton to which it is bound and not disturbed by collisions with other atomic particles, will spontaneously change its spin to the opposite sense. This change drops the atom into a lower-energy state, creating a 21cm photon, which is one whose wavelength is 21cm, to carry away the difference in energy. Within an interstellar cloud an electron may actually reverse its spin much sooner, as often as once every several hundred years, during collision with a passing atom. Random collisions between particles in the interstellar medium can also transfer kinetic energy to a bound electron and cause it to flip over and align its spin with that of the proton.

Even though the time lag for producing a 21cm photon is inordinately long, a ready supply of 21cm radiation is always available because of the enormous number of hydrogen atoms along a line of sight through the Galaxy.

The emission of 21cm photons not only confirms the importance of hydrogen as the primary constituent of the interstellar medium but also provides radio astronomers with a valuable tool for studying the structure of the interstellar medium throughout our Galaxy and neighboring galaxies. Because of its long wavelength, a 21cm photon can travel greater distances through interstellar space than can photons of visible light. This is so because electromagnetic waves are more likely to interact with bits of matter the closer their wavelengths are in size to the characteristic size of the matter. Thus electromagnetic waves with visible wavelengths interact most with the dust, than do electromagnetic waves with very long radio wavelengths.

Since 1963 radio astronomers have found a surprising number of interstellar molecules, including many organic ones, that is, those containing carbon, by searching for their spectral fingerprints, which are emission lines that occur in the centimeter and millimeter regions of the electromagnetic spectrum. From approximately 150 radio spectral lines over 50 molecules have been identified, containing mostly combinations of hydrogen, carbon, nitrogen, and oxygen. The number of different molecules discovered continues to increase. Some of the discovered molecules are familiar inorganic compounds, such as ammonia, water, and several containing sulfur.

A number of interesting organic molecules have also been found, such as formaldehyde, methyl alcohol, and ethyl alcohol. Enough have been found of these interstellar molecules that contain a reasonably complicated arrangement and number of atoms to suggest that, however they are produced, the process is quite capable of forming rather complex molecules. Some of the organic interstellar molecules have not yet been produced in a chemistry laboratory, so that study of the interstellar medium is adding a new dimension to organic chemistry.

Compared with hydrogen, the amounts of other molecules that occur in interstellar space are small, less than one-thousandth that of hydrogen. A few of their spectral lines are observed as absorption lines instead of emission lines whenever enough molecules lie along the line of sight toward a Galactic or extragalactic radio source that emits continuous radiation. Molecules are primarily found in dark cloud complexes such as those in the constellations Orion, Taurus, and Ophiuchus. Other locations for molecules are distributed across the Galaxy in localized regions containing interstellar clouds. Some are even concentrated in tiny high-density regions comparable in size to the Solar System. In addition to being found in our Galaxy, the hydroxyl radical, water, formaldehyde, hydrogen cyanide, ammonia, and carbon monoxide have also been detected in several nearby galaxies. Thus their presence in the interstellar medium of our Galaxy is not a unique event, but probably represents a common feature of most galaxies. When we discuss the structure of our galaxy, we will have more to say about the locations of molecules.

How these molecules were formed is not well understood. A two-atom collision can produce a diatomic molecule, but it is very difficult to imagine a sequence of successive collisions that can produce a polyatomic molecule with as many as 13 atoms. But before we discuss a possible formation mechanism for the molecules, we should introduce interstellar dust.

Interstellar dust consists of solid grains of microscopic size whose composition and properties are very unlike most types of dust on Earth. Photographs of

217

regions along the Milky Way are laced with dark patches that are large clouds containing dust as well as gas. The dimming of starlight is caused almost entirely by interstellar dust, for the gaseous component of interstellar matter is reasonably transparent to starlight. In fact, the interstellar gas is billions of times more transparent to visible light than is air at sea level on Earth. Clinging close to the Galactic plane, within a few hundred light years, interstellar dust completely shuts off our view of the Galactic center in visible wavelengths, and it keeps us from seeing extragalactic objects whose direction is along the Galactic plane. In other spiral galaxies seen edgewise this dust is the dark lane that passes centrally across the galaxy.

Dimming by interstellar dust is greatest for ultraviolet light, less for visible light and infrared wavelengths, and least for radio wavelengths. For visible light, the loss can be as much as 0.7 magnitude per 1000ly (the average is about half this value) near the Galactic plane. This means that for a star at the center of the Galaxy, about 30,000ly away, only about 1 photon out of every 100 billion photons reaches us. If we do not correct the observed apparent magnitude of a distant star for this loss of light, its distance calculated from the distance modulus is too large. In the hard X-ray, infrared, and radio spectral regions, however, we can observe all the way to the Galactic center, since the dust is less absorbing at these wavelengths. Because blue light is affected twice as much as red light by interstellar dust, light from a distant star not only looks dimmer but is also redder than it should be for the spectral type of the star. Astronomers refer to this effect as interstellar reddening. Because of this effect, color indices measured for distant stars are in error and must be corrected before they can be used as a measure of the star's temperature.

About 1% of the mass of interstellar clouds is due to dust and 99% is due to gas. The average density of the dust is about one grain per 10^{13}cm^3, or one grain in a cube 200m on each side. This is a very low density when compared with the typical interstellar gas density of one atom per cubic centimeter. The density of the dust grains can be much larger in small, localized regions, such as the heart of an interstellar cloud. But so will the density of gas also be larger, and it appears that the ratio of dust to gas in most of the interstellar medium is constant to within a factor of about 2.

What is the size and composition of dust grains? Because the scattering of photons is strongly dependent on the size of the scattering particle, the strong scattering of visible light by interstellar grains suggests that a grain is comparable in size to the wavelength of light, say from a few Angstroms to about 0.3 micrometers. At this size the typical grain should contain about 100 million atoms with most of them being elements heavier than hydrogen and helium. Initial analyses of data from the Infrared Astronomical Satellite suggested that there are diffuse dust grains lying in between interstellar clouds that are less than 100 Angstroms in size. From the reddening, dimming, and polarizing of starlight, astronomers conclude that dust grains, whether large or small, are probably assorted carbonaceous dust and silicates in dense clouds. The exact composition is still uncertain.

It seems possible that most of the interstellar dust comes from material that is being blown out of stellar atmospheres. There are several phases in a star's life, starting from birth and ending with death, when the star can lose matter. As an example, among the brightest sources of infrared radiation are the glowing dust shells around some stars, called circumstellar shells. Apparently, the grains intercept short-wavelength radiation from the central star, heat up, and reradiate the energy as long-wavelength infrared photons.

Interstellar dust grains may be the means of forming interstellar molecules. It is thought that hydrogen and other types of atoms can accrete on the cold surfaces of the grains, where they bond together to form molecules. These molecules can escape from the grain surface by absorbing a low-energy photon of starlight, or by some other means, such as the binding energy. Apparently, the enveloping dust cloud prevents ultraviolet starlight or other energetic photons from reaching the interstellar molecules deep inside interstellar clouds and dissociating them.

9.4.   INTERSTELLAR CLOUDS

In photographs of the Milky Way, our view of the starry background is partly or wholly blocked by dark interstellar clouds, sometimes called dark nebulae. They contain denser concentrations of interstellar dust than occur generally in the Galactic plane. One such dark region is a long, chainlike complex composed of dozens of isolated and connected dark interstellar clouds that stretches about halfway around the Milky Way from the constellation Cygnus to Crux. This obscuring strip forms the Great Rift dividing the Milky Way into two branches. In many regions along its length this dark nebulosity separates into tangled lanes of absorbing material that partially cover bright, glowing, gaseous nebulae.

Even though ground-based observations have provided us with important information about the properties of interstellar clouds, much of our understanding of them has come from ultraviolet studies, such as those first done by the Copernicus and IUE satellites. We find that the clouds can be divided into diffuse clouds, which are thin enough for us to observe stars behind them, and dark clouds, which are so opaque that stars behind them cannot be seen. Some of the dark clouds are of immense extent and are the locations of many different types of molecules; these are known to astronomers as giant molecular clouds. The intercloud region (that is, the space between clouds) contains a high-temperature, low-density gas, much of it ionized hydrogen, and wispy-structured dust regions.

For astronomers it is still not clear in all the examples under study where individual clouds leave off and groups of clouds (or cloud complexes) begin. This probably accounts for some of the wide variations in properties quoted for interstellar clouds. Both types of clouds are irregularly shaped and are from 0.1 to 50ly in diameter. Giant molecular clouds can be as much as several hundred light years across. Their temperatures go from about 100K for diffuse clouds down to 10 to 20K for dark clouds. Interstellar clouds may take up as much as 4% of the space in the Galactic plane, with typical masses of several solar masses up to 10^4 solar masses for diffuse clouds and up to 5×10^5 solar masses for giant molecular clouds. Their densities, which vary from about 10 particles per cubic centimeter for diffuse clouds to more than 1 million particles per cubic centimeter for dark clouds, are low compared with the 10^{19} molecules per cubic centimeter in the air we breathe. Even so, dark clouds can be remarkably opaque because of the accumulative effect of extinction by interstellar dust as starlight traverses their enormous lengths.

Typical separations between clouds appear to be on the order of hundreds of light years. The total number of giant molecular clouds in our Galaxy may run up to several thousand, representing a couple of billion solar masses of interstellar matter. The largest single concentration of giant molecular clouds is a ring of them, laying some 15,000ly from the center of the galaxy. It has been suggested that this ring may contain as much as 90% of all the interstellar matter in our Galaxy.

Interstellar dust, concentrated in interstellar clouds in the plane of our Galaxy, obscures distant portions of the disk from our view. The center of the Galaxy in the direction toward Sagittarius is completely hidden in visible wavelengths. Light from distant galaxies that lie in a direction vertical to the Galactic plane is dimmed by 40%. The dimming becomes even greater in directions farther from the vertical until it is nearly total along the plane of the Galaxy, especially toward the Galactic center. Because interstellar dust tends to concentrate in clouds, the dimming of light by it is not uniform but depends on our line of sight through the scattered interstellar clouds. In addition to dimming the light passing through it, interstellar matter also reddens the light by scattering blue-wavelength photons more proficiently than it does red-wavelength photons.

In dark interstellar clouds, hydrogen is primarily in the form of molecules rather than atoms, so that clouds are not sources of 21cm radiation. Unfortunately, molecular hydrogen has no spectral features in the visible or near-infrared part of the spectrum. With the discovery of strong emission in the radio spectrum due to carbon monoxide, radio astronomers have acquired a marker of molecular hydrogen's location and a new probe for investigating dark clouds. Carbon monoxide (CO) can serve as a marker since the conditions that permit it to exist are also suitable for the existence of molecular hydrogen. Dark clouds are the primary locations for interstellar molecules, and the CO molecule is much more abundant in them than in the interstellar medium generally. For every 10,000 hydrogen molecules in clouds, there is approximately 1 CO molecule. Not only can astronomers determine the motions of the molecules from Doppler shifts, but they can also infer densities and temperatures.

Some of the dark interstellar clouds are detectable from their continuous emission of radio waves and infrared radiation. These clouds obviously have sources of energy within them. In 1965, astronomers accidentally found microwave emissions produced by hydroxyl radicals coming from dark clouds. The character of the emissions was peculiar, and the region from which they came was very small. They found that these small regions were also bright sources of infrared radiation, but that they emit virtually no visible light.

The radio emissions were much stronger than could be accounted for by random thermal collisions. Clearly there was a mechanism that selectively excited the hydroxyl (OH) molecule. It is thought that infrared radiation from nearby stars excites OH molecules and they are stimulated to de-excite by interaction with stellar photons having the right wavelength. The emitted radiation in turn stimulates other molecules to radiate in the same fashion, producing an avalanche of emissions. Thus the radiation in a normally weak line is greatly amplified. The word maser used to describe this phenomenon is an acronym for microwave amplification by stimulated emission of radiation.

Astronomers know of several hundred OH masers and several dozen H_2O masers operating in dark interstellar clouds. They are also found in the atmospheres of red giants that are variable stars. In general, masers in molecular clouds are brighter than those in luminous red stars, but those associated with stars seem to be more numerous. Our interest here, however, is the significance of the presence of masers in clouds where astronomers believe stars are forming. Clearly the masers in clouds signify that energetic events are occurring at specific points in molecular clouds. Such events most likely constitute star formation.

9.5. EMISSION NEBULAE

The emission of ultraviolet photons by O and B stars is so great that even far from these hot stars the number of photons is sufficient to ionize hydrogen gas in interstellar clouds. With the ionization of hydrogen, the HI region becomes an HII region, or an emission nebula. Stars of spectral type O5 emit enough ultraviolet photons to ionize hydrogen out to distances of 300ly from the star. For cooler spectral types the surrounding HII region is smaller; an A0 star creates an ionized region about it that is less than 1ly in radius. Emission nebulae are among the most beautiful of all astronomical objects.

In panoramic photographs of the plane of the Milky Way, one sees many bright, glowing regions whose spectrum is an emission spectrum. The Balmer alpha line of hydrogen is responsible for the vivid red color of many HII regions. These HII regions are produced by hot stars and are associated with interstellar clouds either by being surrounded by them or by being on the edge of a cloud complex.

HII regions occur in about six distinct categories, depending on their size and the density of free electrons resulting from the ionization of hydrogen. Astronomers refer to the smallest as ultra-compact HII regions and the largest as supergiant HII regions. The smallest ones are from a few tenths to a few tens of light years in diameter and their masses range from a few tenths to a few solar masses. These smaller HII regions are generally buried in dark molecular clouds so that in the visible part of the spectrum they are almost totally obscured from view or are heavily reddened, if at all visible, the larger ones generally can be seen.

In the three categories of large HII regions, the regions are from a few to 1000ly in diameter, and they contain anywhere from tens to hundreds of millions of solar masses of ionized matter. These three categories are the types of emission nebulae that are seen most readily, with the supergiant HII regions being by far the brightest objects in the spiral arms of our Galaxy and other spiral galaxies. They occur sometimes in groups and sometimes isolated from each other. As for the shapes of these HII regions, they range from readily definable shapes to large, complex, ill-defined regions. Altogether it is estimated that about 1% of the mass of our Galaxy is tied up in the form of HII emission nebulae.

Carina and other gaseous nebulae are suffused with an X-ray glow resulting from many supernova outbursts. In the constellation Cygnus, lying about 7000ly from the Sun, beyond the bright star Deneb, and partially hidden behind the dark interstellar cloud complex known as the Great Rift is a rarefaction in the interstellar medium known as a super-bubble. It is about 1000ly in diameter and contains gas at temperatures of about 2 million K. It appears that this super-bubble was created by a chain of supernova explosions and possibly amplified by stellar winds occurring within the last 3 million years. Such bubbles, surrounding many stellar associations of massive stars, occupy at least as much as 10%, if not more, of the entire Galactic disk and thus are important components of the Galaxy.

Before leaving our discussion of the interstellar medium, we should ask about the nature of the interstellar matter that surrounds our Solar System. From what we have said about giant molecular clouds, the Sun is obviously not sitting in the middle of one of them. Observations with the ultraviolet satellites place the Sun in the low-density, about 0.1 particles per cubic centimeter, and high-temperature gas of the inter-cloud region. Also, our Solar System seems to be located on the edge of a "hole" or a bubble, smaller than a super-bubble, in the interstellar medium that may well be the result of a supernova outburst. Although the Sun in its motion relative to the stars of the solar neighborhood could encounter a dense cloud, greater than 100 particles per cubic centimeter, it is not likely to happen soon.

9.6. REFERENCES

Verschuur, G.L. (1988). Interstellar Matters: Essays on
    Curiosity and Astronomical Discovery. Springer-
    Verlag: Berlin.
Wynn-Williams, Gareth (1992). The Fullness of Space:
    Nebulae, Stardust, and the Interstellar Medium.
    Cambridge University Press: London, England.
Whittet, D.C.B. (2002). Dust in the Galactic Environment
    (Series in Astronomy and Astrophysics). CRC
    Press: London.

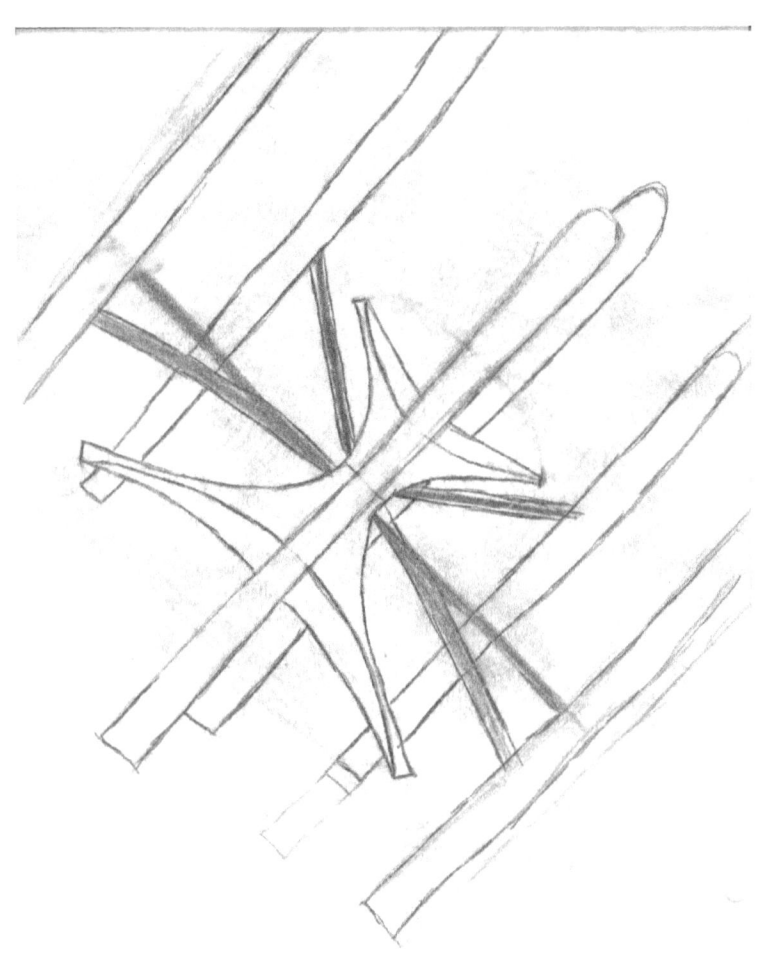

227

Chapter 10: The Theory of Relativity
By John C. Evans and Harold A. Geller

The Newtonian theory of gravity was very successful in explaining most gravitational problems involving the material objects of the macroscopic world. But for the motion of bodies at high velocities or moving in gravitational fields of very massive and compact bodies, Newton's concepts of motion and gravity yield incorrect answers. In the early part of the 20th century, Einstein's relativity theory provided correct answers to problems on which Newtonian ideas failed, but more important, Einstein's ideas produced a profound change in our understanding of the physical world.

10.1. EINSTEIN'S RELATIVITY

So, what is relative in relativity theory? To answer this question, consider a simple example. If we walk down the aisle of a moving train car, the car itself represents a frame of reference for describing our motion. Looking out the train window we see our immediate frame of reference, the train car, moving relative to another frame of reference, the Earth's surface. We know that the frame of reference provided by the Earth's surface moves relative to yet another frame of reference, the Earth's center, about which the Earth rotates. There is still another frame of reference, the Sun, from which to view our stroll down the aisle of the train, since the center of the Earth moves relative to it. This continues to even larger scales in the Universe. We cannot ascertain clearly an absolute motion in the Universe since we can find no frame of reference that is absolutely still. This is but the first of what will be several truths that seemingly contradict our intuition. All motion is relative; there is no absolute standard of rest.

Newtonian described space with three spatial

coordinates that describe where an object is located in a frame of reference. Location does not necessarily have anything to do with time. Newton believed space existed "without relation to anything external" and time "flows equably without relation to anything external." In Newton's view space fills the Universe like a rigid framework while some universal clock ticks relentlessly onward. To Newton an absolute space and time existed independent of the material Universe, and could be thought of as preceding the material Universe's existence and surviving its demise.

Einstein found Newton's concept of an absolute space and time unacceptable. Einstein knew that we can neither drop anchor nor become external observers watching passively what goes on throughout the Universe. Each of us is both observer and participant. The special theory of relativity of 1905 incorporates this concept, and it deals with motion in which gravitational effects are not involved. As basic postulates, Einstein assumed that the laws of physics must be the same for all observers in uniformly moving frames of reference, and the velocity of light is the same for all observers. By 1916, Einstein had worked out another, more comprehensive theory, which was an alternative to Newton's gravitational theory, called General Relativity (GR). This theory describes non-uniform, or accelerated, motion, where the special theory does not.

In Einstein's general theory, he conceived of space and time as inextricably bound together, so that the three dimensions of space are coupled with that of time. Einstein describes gravity in terms of the geometry of this four-dimensional space and time. In the vicinity of material bodies, space and time possess a local curvature; the more massive is a body, the stronger is this curvature. Where no massive body exists, the local geometry of space and time is flat, that is, it shows no curvature.

In the nineteenth century, space was thought to be an absolute frame of reference relative to which absolute motions could be measured. Space was also imagined to be filled with a stationary, invisible medium, called aether, that carried electromagnetic waves (as air transports sound waves). If light must move through such a medium, it was argued that its speed should differ depending on its direction relative to a moving observer. Thus it was felt that Earth's absolute velocity should be measurable against this stationary medium by timing the speed of light in various directions.

In 1887, the American physicists Albert Michelson (1852-1931) and Edward Morley (1838-1923) sought to detect a difference in the speed of light beams propagated over the same distance, one parallel to Earth's motion around the Sun and one perpendicular to it. The time light spent moving perpendicular to Earth's path was calculated to be $1-(v^2/c^2)$ times shorter than the time light took to move parallel to Earth's path, where v is Earth's velocity and c is the velocity of light. But no matter which direction measurements were made or at what time of year the experiment was performed, the result never changed. However, there was no measurable difference in the times. This seemingly meant that one could not measure Earth's absolute velocity in space.

The Michelson-Morley experiments were bewildering to the scientific community at the time Einstein began work on his special theory. Einstein rejected the notion of aether and concluded that the solution to the problem lay in recognizing the inadequacy of the Newtonian concepts of space and time. To develop the mathematical formulation of his theory, he laid down two postulates. First, the laws of physics are the same for all observers in uniformly moving reference frames. Second, the velocity of light, 299,792km/s, is the same for all observers in space regardless of the motion of the source or observer.

As an example of the consequences of the first postulate, suppose that two observers are moving uniformly relative to each other (not accelerating). They are said to occupy inertial (or non-accelerating) frames of reference. According to the first postulate, neither reference frame is preferred by the laws of physics; there is no way to distinguish one from the other. For that matter, all frames of reference moving uniformly relative to one another are indistinguishable. Consequently, neither observer is able to determine the uniform motion of his or her reference frame by any experiments conducted within that frame of reference. For example, a person occupying an inside cabin (no portholes) of a ship cruising at uniform speed in calm seas cannot conduct any kind of experiment to indicate whether the ship is moving uniformly at sea or still tied to the dock. That is, experiments like throwing a ball, flipping a coin, or playing a game of billiards do not distinguish between the two situations. If the person performs another experiment, using a frame of reference different from that provided by the cabin, say, by looking out a porthole at the ocean, he or she can determine the state of motion relative to the other frame of reference. But excluding that possibility, the only means of detecting motion is when it is accelerated motion, such as when the ocean is stormy.

Einstein's second postulate is obviously consistent with the Michelson-Morley experiment. The speed of the light beam in the experiments did not depend on the direction of travel; it did not depend on the light source; it was not affected by Earth's motion. In other words, the speed of light is independent of the relative velocity of source and observer. Regardless of whether we measure the velocity of light in our reference frame or in some other, we find that the velocity of light always has the same value. Einstein reasoned that what is different between frames of reference moving relative to each other is the measurement of space and time.

Since light brings us information about the Universe and it travels at a finite rate, albeit a very large one, we are actually looking back in time when we look out into space. The nearer the object, the shorter the look back in time, but the objects we see farthest away are objects we are seeing as they were longest ago. Thus the night sky alone demonstrates the coupling of space and time.

Let us look at some further consequences of Einstein's postulates regarding the measurement of space and time and, in addition, mass. Suppose two observers A and B are in different reference frames that are at rest relative to each other. If both make measurements of lengths in space and intervals of time, they should naturally agree on the measurements; consequently, they must occupy the same realm of space and time. Now suppose the reference frame containing B moves off at a large fraction of the speed of light relative to the frame containing A. Do the two observers still occupy the same realm of space-time as before?

On the basis of his two postulates Einstein derived three important formulas for length, mass, and time in which the factor $1-(v^2/c^2)$, called the Lorentz contraction factor, plays a crucial role. The Lorentz contraction factor is named after the Dutch physicist Hendrik Lorentz (1853-1928), who derived it in 1904 from his mathematical analysis of electromagnetism. The length, mass, and time interval for an object, say, a clock, in B's reference frame are measured by observer A, who sees the object in motion. If A has an identical clock with him, then his measurements of B's clock are related by the Lorentz contraction factor to his identical clock's length, mass, and time interval. These effects are unimportant for ordinary velocities which are very much less than the speed of light, so that A and B still occupy approximately the same realm of space-time. But this is not the case for speeds comparable to the velocity of light.

What do the formulas tell us? Observer A, who sees the object moving with B's reference frame, will measure a shorter length for the object in the direction in which it is moving than will observer B, to whom the object is stationary. For example, observer A would notice that B's rocket ship is also contracted in length compared with his own rocket ship, which is, of course, his frame of reference. Thus there is a contraction of space in the direction of motion for the moving frame of reference as seen by A. Observer A will measure a greater mass for the object in B's moving spaceship than observer B will, and observer A will also measure a longer time between two events, such as the tick of the clock taking place on B's rocket ship, than observer B will for the same two events. For example, an event lasting 1s on B's clock is stretched out to 5s as A sees B's clock if B is moving at 98% the velocity of light relative to A. The latter effect is called time dilation, or spreading out of time. The faster observer B travels relative to observer A, the slower his clock appears to run as observer A sees it (observer A's own clock is seen by observer B also to be running slow). Neither observer sees any effect on his own clock, which is therefore said to measure "proper time" for each respective observer. Thus measurements of length, mass, and time vary with the frame of reference. Clearly, observer A and observer B occupy different realms of space-time.

From the preceding discussion, what can we infer about space and time for the Universe as a whole? Newton's concept of an absolute space and time envisions a material Universe inserted into preexisting space and time. But in Einstein's concept, space and time are in the Universe; that is, the Universe creates space and time in both a local sense and a global sense. That is, there is no space beyond the Universe, and there is neither time before nor after the Universe. Space and time and their local features are properties of the Universe.

The concepts of relativity theory seem at first contemplation to be contrary to ordinary experience and to so violate common sense that cannot be of any consequence. This is far from the case, for relativity theory has replaced old ideas of space and time with a unified theory that does indeed encompass common experience and at the same time leads us to new and unsuspected revelations about physical reality.

In 1916, Einstein advanced his theory of relativity greatly by making it apply to observers (reference frames) moving nonuniformly (accelerating) relative to each other. Nature's fundamental laws, he reasoned, remain invariant throughout the Universe in all frames of reference, whether the observers are accelerated or not.

In his second law of motion Newton had shown that the force it takes to accelerate a body is proportional to its inertial mass. Inertia is the resistance a body offers to an applied force. He was also aware that the gravitational force on a body is proportional to its gravitational mass. Otherwise, bodies of different masses would not fall to the ground at the same rate, that is, with constant acceleration, as we know they do. In 1889, a Hungarian physicist, Baron Roland von Eotvos (1848-1919), first proved experimentally and very precisely that inertial and gravitational mass are equivalent, an equality that had long been taken for granted. Modern experiments confirm that the two masses are the same to at least 1 part in 10^{12}.

Einstein argued that the equality of inertial and gravitational mass must mean that "the same quality of a body manifests itself according to circumstances as `inertia' or as `weight.'" The consequence of this is that it is impossible to distinguish between the effect of an inertial force or a gravitational force on accelerated motion. Einstein worked the idea into the Principle of Equivalence, which states that a gravitational force can be replaced by an inertial force that is due to accelerated motion without any change in the physical activity.

By way of illustration, imagine an observer in a

rocket ship whose acceleration is that which he or she would feel at the Earth's surface, or a constant 1g. Because 1g is the acceleration we experience on Earth's surface, the observer should experience within his or her reference frame, defined by the cabin of the rocket ship, what he or she would experience on Earth. To show this, suppose the observer drops a ball. The ball continues to move forward with the velocity the ship had at the moment it was released. If the ship were moving forward at constant velocity, the ball would remain suspended at the same point because ship and ball move the same amount. But the ship is accelerating, so the floor moves forward faster than the ball does, ultimately colliding with the ball. The observer in the ship could attribute this to Earth's gravitational attraction, if he or she thought the ship were not accelerating but was still sitting on the surface of the Earth.

Einstein pointed out that the principle of equivalence gives an observer alternative descriptions of events in that he or she can replace the force of gravity by an inertial force caused by accelerated motion. An inertial force is not a real force but an effect of non-uniform motion by the observer's frame of reference. You have already experienced such a fictitious force from accelerated motion when standing on the floor of a merry-go-round. You felt a force called a centrifugal force that tended to move you toward the rim.

In GR, spatial curvature of local space-time is determined by the masses of local bodies. If no mass is there, the curvature of nearby space is zero, and it is a flat space whose geometrical properties are described by ordinary Euclidean geometry (the kind you learned in high school). In the warped geometry of space-time that surrounds a large mass, less massive objects move along curved paths. An example is a planet's elliptical path around the Sun. The planet moves in a curved path in the warped space surrounding the Sun. The only way we have of illustrating curved space for you is in two dimensions rather than three.

The force we call "gravity," then, is nothing more than the natural behavior by bodies moving within the geometrical framework of space-time. Newtonian physics says that the body moves according to action from a distance dictated by a force called gravity, whereas relativity says that a body follows the shortest available path in response to the local structure of curved space-time.

Massive bodies, in addition to warping space in their vicinities, also alter time. If we could position ourselves where there is little if any warping of space-time, far from a massive body, and watch what happens as a clock approaches the massive body, we would see the clock ticking slower and slower the closer it gets. Clearly the realm of space-time in the immediate vicinity of massive bodies is different from that far from any mass. We may summarize motion with the principle of GR, which states that curved space-time tells matter how to move, and in turn matter tells local space-time how to curve.

With some exceptions, the realms of space-time in which we find various astronomical objects are approximately the same as ours. Thus Newton's laws of motion and law of gravitation describe what is happening for these bodies. However, we must face the cosmic reality of different realms of space-time when we discuss black holes.

10.2. TESTING RELATIVITY

After Einstein presented his relativity to the scientific community, he and others looked for observable consequences of the theory. Three were recognized quickly, which are now referred to as the classic tests of relativity: (1) the bending of starlight; (2) the advance of Mercury's perihelion; and, (3) the gravitational redshift.

The first to actually be observed was the deflection of starlight passing near the Sun.

Einstein using his GR theory predicted in 1911 that, since space in the Sun's vicinity is curved, a ray of starlight headed toward Earth that passes just outside the limb of the Sun will follow a curved path in the curved space surrounding the Sun. The ray of starlight will be deflected by a small fraction of an arc second from its original path through space.

How can such a prediction be tested when the Sun is so bright and background stars so faint that they are not visible near the edge of the Sun? During a total solar eclipse bright stars are visible on photographs of the darkened sky around the eclipsed Sun. Another photograph of the same area of the sky can be made at night, a few months earlier or later, with the same telescope when the Sun is in a different place in the sky. Star positions on the two photographs can then be compared and will show that the stars around the eclipsed Sun have shifted away from the Sun. The amount of deflection decreases with the star's distance from the Sun's limb.

This experiment was carried out for the May 29, 1919, total eclipse by two teams of British scientists, one lead by Arthur Eddington (1882-1944). Their results and numerous measurements since all support the predictions of Einstein's GR and confirm the existence of a region of curved space surrounding the Sun. The success of his theory projected Einstein into the public limelight from which he never disappeared. Until his death in 1955, Einstein remained somewhat of a folk hero because of his intellect and humanity to the common people of the world, and honor that essentially no other scientist has ever achieved.

Einstein's theory of relativity cleared up a long-standing problem that had plagued astronomers for decades. This was the moving of the perihelion point (or rotation of the major axis) of Mercury's orbit in the direction of the planet's revolution around the Sun, which is called the precession of Mercury's orbit. Although attractive forces of gravity, technically called perturbations, from other planets are the primary cause, astronomers observed that the orbit was precessing by an additional 43 arc seconds per century more than Newtonian theory could explain. To account for the additional 43 seconds, they faced the unwanted choice of either increasing the mass of Venus by an inadmissible one-seventh or postulating the existence of a never-observed planet called Vulcan within Mercury's orbit. Einstein's theory of GR removed the difficulty.

GR accounts for, as does Newtonian theory, the rotation of a planet's orbit in its own orbital plane. In either theory, the change in perihelion is most pronounced for Mercury's orbit because it is closest to the Sun and has the most eccentric orbit. But Einstein's equations for the elliptic motion of a planet about the Sun include a term not present in the Newtonian equations. Its contribution is a tiny fraction, the 43 arc seconds per century, of the total orbital precession. This additional 43 arc seconds adds up to one extra revolution in 3 million years. Looked at relativistically, the planet's eccentric orbital motion periodically moves it into stronger and weaker gravitational fields, where it encounters a different space-time curvature.

In GR Einstein showed that, for a distant observer, an object in an intense gravitational field would be contracted, have gained mass, and have slowed down its clock time. Consequently, if the object is an atom, time dilation should also play a role when it emits a photon. The consequence of time dilation in photon emission is that the wavelength of a photon is lengthened, or redshifted, by an amount that depends on the strength of the gravitational field. The value of the change in wavelength depends on the mass of the attracting body divided by its radius.

This effect, known as gravitational redshift, has practical astronomical interest because it occurs when a photon of light escapes from a star. If the star's gravitational field is sufficiently intense, we can measure the change in wavelength. We cannot easily measure this effect for the Sun, but for a white dwarf of solar mass and small size (where the mass divided by the radius is large); the gravitational redshift is a measurable effect. It has been observed for several white dwarfs in binary systems; this is made possible because, by using the companion's spectrum, we can differentiate between a gravitational redshift of the spectral lines and a Doppler shift produced by the system's radial velocity. The measured redshifts agree satisfactorily with those predicted by Einstein's theory. Finally, the gravitational redshift has been verified with even greater accuracy in a laboratory experiment.

One of Einstein's predictions lay idle for nearly half a century as too difficult to verify. That is, extremely weak gravitational waves are radiated into space with the velocity of light by rapidly accelerated or orbiting bodies. Gravity waves are ripples in the overall geometry of space and time.

For example, two stars in a binary system continually alter the geometry of space and time in their vicinity because of their orbital motion. These adjustments appear as tiny ripples that propagate out through all of space and time like ripples on the surface of a pond. Gravity waves might be detectable with a very sensitive apparatus for large astronomical objects undergoing violent activity, such as supernova outbursts or the nucleus of an active galaxy, but unfortunately not for small events like ordinary binary stars. The detecting apparatus works on the principle that any gravitational wave passing through the detector momentarily deforms space and causes the detector to vibrate slightly.

Many years ago experiments tried to pick up (inside

large, suspended aluminum cylinders) infinitesimal oscillations that would be produced by gravity waves rippling the space occupied by the cylinders. At first it seemed that they had succeeded in detecting gravity waves coming from the Galaxy's center. But so far, other and far more sensitive gravity-wave detectors, that can detect deformations as small as 10^{-17} cm, have failed to find any evidence for gravitational radiation.

However, indirect evidence of gravitational radiation has been found in the radio observations of binary pulsars. The gravitational interaction between the pulsar and its close companion, perhaps a neutron star or white dwarf, results in part of the orbital kinetic energy being radiated away in the form of gravity waves. The loss in energy decreases the orbital separation between the components.

Radio monitoring during the period 1974 to 1979, covering more than 1000 orbital revolutions, showed a decrease in the orbital period of about 101 seconds of arc per year. Allowing for uncertainties in the mass of each component and the inclination of their orbital plane, the result is in reasonable agreement with GR's prediction of 76 seconds-of-arc/year.

10.3. REFERENCES

Goldsmith, D. (1975). *When Time Slows Down*, **Mercury**, May/June.

Hutchings, J.B. (1985). *Observational Evidence for Black Holes*, **American Scientist**, January/February.

Kaufmann, W.J. (1979). **Black Holes and Warped Spacetime**, W.H. Freeman.

Chapter 11: The Quest for Life
By Harold A. Geller and John C. Evans

During the course of the history of the Universe, life became possible only after galaxies had formed and their stars had existed long enough to synthesize the heavy atoms found in organisms. The six most important elements for life on Earth, and their percentages by numbers of atoms in human beings are as follows: 61% hydrogen; 26% oxygen; 11% carbon; 2% nitrogen; and, less than 1% phosphorus and sulfur. If the primordial condensate that immediately followed the big bang had not cooled rapidly in the first few minutes, life could not have occurred in the Universe because hydrogen would mostly have been converted into helium, leaving little available for stellar nuclear synthesis, and heavy elements would not have been subsequently formed in stellar interiors.

Thus an important point is whether enough time has elapsed for stellar evolution to build a significant abundance of the biologically significant atoms from which chemical evolution will flow if given the opportunity. Clearly it has at our position in the Galaxy, or we would not exist.

Biologists are not unanimous on all the factors in the definition of life, but most agree that life is different from nonlife. Life evolves or changes, by chance mutations or otherwise, as time goes on while interacting with its environment in a unique way. Furthermore, life is not a "thing," it is a process made up of an unimaginably large number of complex chemical reactions that collectively produce all the characteristics we associate with life.

11.1. LIFE IN THE SOLAR SYSTEM

The place to begin our discussion is with ourselves, since we are the only system of life for which we have any information. From the Earth we can then move to consider the Solar System; and in the next section, we shall consider the question of life beyond the Solar System.

The thermal habitable zone in our Solar System, that is, the region where life might most likely flourish on a Terrestrial planet because of the presence of liquid water, between 273K and 373K, lies between Venus and Mars. The inner limit is roughly the point where water would boil, and the outer limit is the point where water would freeze. Had Earth formed somewhat closer to the Sun than it did, the greenhouse effect might have become dominant, as it is on Venus, resulting in a hot, sterile surface. And had the Earth formed farther away from the Sun than it did, water would have remained frozen as subsurface ice or polar caps, and the surface of the Earth would somewhat resemble the cold Martian landscape.

Our Earth, therefore, is at the right distance from the Sun for development of an active biosphere. The chemistry of Earth's life is built on the chemistry of the element carbon and the solvent water, supplemented by the biologically important atoms hydrogen, nitrogen, oxygen, phosphorus, and sulfur. Carbon atoms bond easily with other carbon atoms, producing long chains to which other biologically significant atoms can bond. The resulting molecules, whether or not they are part of or a product of living matter, are called organic molecules.

Water, because it can flow readily and remain in liquid form through a range of conditions, is the ideal solvent for many organic compounds. From water come hydrogen bonds, which give structural stability to strings of proteins, nucleic acids, and other long-chain carbon compounds. A liquid water environment and moderate temperature make it possible for such long-chain carbon molecules to form, and they have become the biochemical basis of life as we know it.

Even with this biochemical basis, life would not have

been able to develop and sustain itself without proper temperature, a supply of nutrients, self-regulating mechanisms, and the Sun's energy. The Sun is the prime source of energy for driving the chemical reactions in the cycle of life.

How did life begin, and how did it derive from nonlife? Our contemporary ideas on the evolution of life began in 1924, when the Russian biochemist Alexander Oparin (1894-1980) introduced chemical evolution as a necessary forerunner to biological evolution. In 1928, the English biologist J.B.S. Haldane (1892-1964) independently suggested an outline for chemical evolution, which is still the basis of our current understanding.

Earth's cooling and solidifying crust was racked by volcanic activity that presumably vented carbon dioxide, nitrogen, water vapor, hydrogen compounds, and smaller amounts of other molecules that are easily vaporized. These probably formed Earth's early atmosphere. Today active volcanoes discharge large quantities of carbon dioxide, nitrogen, water vapor, some sulfur, and traces of other gases.

Subsequent cooling of Earth condensed water vapor, forming the warm seas and the shallow lagoons and pools that were destined to provide a haven for the development of organic compounds. From this so-called "primordial soup," over a long time, the more complex organic molecules or biological macromolecules, such as proteins and nucleic acids evolved. These were formed by energy from solar ultraviolet and visible light, electric discharges, and heat from radioactivity, volcanoes, and meteoric impacts.

Cells are the basic units of life from which complex organisms, such as human beings, are built. Modern biology has shown that the cell is in general the minimum organized unit of matter that displays the properties of life. Collections of cells make up an organism.

The cell has in it a water-based substance surrounding a nucleus, which holds coiled, threadlike strands known as chromosomes. The chromosomes, which transfer hereditary characteristics to each generation of new cells, occur in pairs, with a fixed number in every cell of every species. Human cells have 23 pairs. The nucleus, with its many chromosomes, controls the cell's activities. It contains instructions for manufacturing the specialized cells, such as muscle, bone, or liver cells, and maintaining their functions to keep an organism alive.

Hereditary information is contained in the deoxyribonucleic acid (DNA) molecule, a large complex molecule found in the chromosomes of the nucleus of every living cell. One human cell has about 800,000 DNA molecules. Since there are many varieties of living organisms with many chromosomes, there are many different forms of DNA. DNA has two primary functions. It carries the heredity instructions for manufacturing proteins in the cell, and it passes genetic information on to daughter cells during cell division by making copies of itself, which is called replication.

Essentially, the atoms in DNA are linked like a twisted ladder, or a double helix. The spiral is a right-handed spiral in which each tread is of the same size and at the same distance from the next and turns at a rate of about $30°$ between successive treads. The genetic code, which specifies the hereditary message, is carried on the treads of the ladder and is contained in the sequence of the four different units. Thus the language of heredity is written in an alphabet of only four letters. For each protein potentially capable of being formed, a specific segment of the DNA molecule carries the information by which the 20 kinds of amino acid subunits in the protein are properly ordered during its synthesis.

During cell division, replication of the DNA molecule begins with separation of the two spirals. Each strand of the double helix directs the information of a complementary strand to pair with it. They migrate to opposite ends of the cell before the cell divides. Thus in the daughter cells are two daughter DNA molecules identical to those of the parent DNA molecule in the parent cell. DNA molecules specifically determine the type of organism, for example, human being or elephant.

Biological evolution is accomplished by mutations, which introduce new factors into the genetic message of DNA, by recombination, which is the rearrangement or new association of message units by sex-like processes, inheritance from different parents, and by selection, which is the weeding out of inferior traits in a population through successive generations, not in individuals in their lifetimes. DNA molecules are remarkably stable. Mutations occur about once per gene for every 100,000 cell divisions in most mammals. Evolution in plants and animals is controlled by two major forces which are limits that the environment sets for the organisms and changes in their hereditary material. Organisms are constantly modified by chance mutations, sexual selection, and natural selection. Together these produce members of a species that can survive in a changing environment.

From Earth's fossil record two important points are evident. First, species do not repeat, although essential parts of organisms, such as the eye, may have several independent entries into biological populations. And second, some species have remained essentially unchanged since their appearance, whereas others have shown significant change. Finally, an important consideration is whether or not there is a definite direction to biological evolution.

If we assume that chemical evolution is followed by biological life that evolves toward greater complexity, what can we say about the development of intelligence? The great leap forward for human beings came only in the last few million years, when our humanoid ancestors learned to walk upright, freeing hands for the manipulation of tools. As their brains evolved and their mental capacities expanded under pressure for survival, a collective culture and civilization set human beings apart from other living creatures.

Yet human beings, although the most advanced, are not the only intelligent creatures on Earth. One of the most important characteristics of intelligence is the ability to collect and transfer information. There is a spectrum of this ability among Earth's creatures signifying that they possess varying intellectual capacities. Scientists have found a "language" capacity in chimpanzees, gorillas, and dolphins.

It is unlikely that another intelligent creature would closely resemble us. As we have pointed out, the fossil record does not indicate that evolution repeats itself. Loren Eiseley (1907-1977) noted in a more general sense that we might expect along such lines for the Universe at large: "Life, even cellular life, may exist out yonder in the dark. But, high or low in nature, it will not wear the shape of man. That shape is the evolutionary product of strange, long wandering through the attics of the forest roof, and so great are the chances of failure, that nothing precisely and identically human is likely ever to come that way again."

Among meteorites that have been recovered, there is a small subgroup of stony meteorites called carbonaceous chondrites, which contains up to about 5% organic molecules. The fact that chondritic meteorites carry organic compounds has been known for more than a century, but only when techniques had been developed for studying lunar rocks could the organic compounds be definitely ascribed to an extraterrestrial origin. Amino acids were discovered in a fresh chondritic specimen that fell near Murchison, Australia, in September of 1969.

Extraterrestrial amino acids have since been found in other meteorites, but only a few of these acids are in living cells of Earth organisms. Amino acids from meteorites are almost an equal mixture of right- and left-handed molecules, or stereoisomers. Amino acids of biological origin are exclusively left-handed, supporting the idea that the meteoritic amino acids are of extraterrestrial origin. If extraterrestrial life produced amino acids, we believe that they would be exclusively either left- or right-handed, so the mixture of both types in the meteorite specimens suggests a chemical, not a biological, origin.

The Moon was the first body outside the Earth on which searches were made for the presence or remains of living organisms. Organic chemists had improved old techniques and developed new ones in anticipation of new findings on biochemical evolution and the origin of life that might come from the lunar samples from the Apollo program. But, alas, nowhere in any of the samples was a significant amount of carbon found. No amino acids, nor proteins, nor nucleic acids, were ever found. Also, no water molecules, free or chemically bound, were found in these surface materials.

Neither Mercury nor Venus is a very promising site for the development of life. This leaves Mars as the only realistic planet for biological exploration. The Viking Mission to Mars was designed, in part, to search for life on Mars. Two cameras periodically scanned the immediate surroundings to look for any large-scale biological life. But the main biological hunt on Mars was for microorganisms in the Martian soil.

In the various Viking biological experiments, all found an unusually active Martian soil. However, not only was there no other evidence for microorganisms, there was no evidence of the presence of any organic chemicals. It is possible that living organisms developed and evolved in an environment on Mars so different from that on Earth that we did not formulate the experiments correctly, or we are still not interpreting the results properly. Nonetheless, at the present time it appears unlikely that Mars does have or has had living organisms on its surface.

Although evidence exists that suggests extensive chemical evolution has occurred either prior to, or early in the Solar System's existence, whether or not it is heading toward biological evolution anywhere else besides Earth is not evident. At this point it is too early to make a final pronouncement about life in the Solar System.

11.2. LIFE IN THE UNIVERSE

The first question we may wish to ask is about the environment, or platform, on which the slow processes of chemical and biological evolution can take place. An initial thought would be to assume that the life process might begin in a planetary system about a star and on a solid surface of one or more of its planets, as did our own development.

If we decided to select stars in the solar neighborhood near which life might be found, how should we proceed? Some criteria have been proposed that could improve the probability of finding stars with planetary systems on which life might exist. The criteria do not argue against the existence of planets in those situations that are eliminated, but rather they try to identify what are the highest probability situations in which planets can support life.

A first criterion for assigning a low probability to a star's having a viable planetary system is the length of time a star is a main-sequence star, which is the longest phase in a star's life. Thus, the hottest stars would be rejected for life-bearing planetary systems because their time on the main sequence is much too short, less than 1 billion years, to permit chemical and biological evolution, which would probably take several billion years as is the case on Earth. Even if the time scale for biological evolution was assumed to be 1 billion years or less, the quantity of high-energy photons, such as X-rays and ultraviolet, emitted by hot stars would probably prevent the formation of complex organic molecules. Although cool stars on the lower end of the stellar temperature scale have a lengthy life span on the main sequence, they are much too cool to support the development of life except in orbits that would be very near the star and possibly unstable, or at least synchronous orbits. For this reason, they are not assigned a very high probability.

Second, we should probably assign a very low probability to the vast majority of binary and multiple stars, about half the stellar population, because the orbits of planets around them might not be stable enough to maintain a planet in a thermally habitable zone. A third criterion would be how rapidly the host star changes, so that all stars in the post-main-sequence stages of evolution are probably not good candidates. If planetary systems had developed and sustained life, then the expansion of stars during red-giant evolution would destroy the thermal environment in which they had existed. And subsequent phases in a host star's existence in many cases may be too short for life to develop a second time somewhere else in that system.

Such arguments leave us with main-sequence stars with surface temperatures from 3700 to 7500K as the best possibilities for supporting life. If there are approximately 400 billion stars in our Galaxy and 85% are main-sequence stars, then there are about 340 billion main-sequence stars. Astronomers estimate that about 90 percent of main-sequence stars have a surface temperature from 3000 to 7500K, or what amounts to about 300 billion stars. One-third of these, or 100 billion, are apparently from 3700 to 7500K, and one-half of these, or 50 billion, are probably not members of a binary system. Thus there are a large potential number of life-supporting planetary systems. In fact, estimates from the Kepler mission, specifically designed to detect extrasolar planets around distant stars in a specific region of the galaxy, place this estimate as being closer to 100 billion, nevertheless, the same order of magnitude.

To define a habitable zone for life, why do we seek a temperature range in which water is in liquid form? Why not some other organic solvent, such as alcohol or ammonia? First, water is a simple molecule, consisting of just three atoms, of which two, hydrogen, are the most abundant element in the Universe, and the third atom, oxygen, is among the most abundant elements after hydrogen and helium. Second, liquid water can store a great deal of thermal energy before it vaporizes. Thus it acts as a buffer to day-night temperature changes that occur when a planet rotates. Finally, water has a high surface tension, which can help concentrate solids at its boundaries. In a similar vein, carbon chemistry is expected to be a more widespread basis for life than is silicon chemistry or germanium chemistry, both silicon and germanium behave chemically somewhat like carbon. This is so because carbon is far more abundant cosmically than either silicon or germanium.

Another is whether chemical evolution up to the macromolecule stage will be followed by biological evolution, given a suitable environment and sufficient time. Although biological evolution leading to simple life forms is obviously less probable than chemical evolution is, it seems reasonable that biological evolution has occurred many times just from the huge number of opportunities.

Of the 31 stellar systems within 15 light years of the Sun only 3 seem to meet the criteria needed for what some call an ecoshell. These are the main-sequence stars Epsilon Eridani, Epsilon Indi, and Tau Ceti. If we take the solar neighborhood as a representative sample, in which 3 stars out of the 31 stellar systems within 15 light years of the Sun have potentially habitable planets, the average distance between biologically suitable stars is about 17 light years.

Therefore, within a radius of 1000 light years, then, we should expect to find somewhat less than 1 million stars having suitable planets harboring some kind of life. Even if only 1 in 1000 of these planetary systems has an intelligent species, that still leaves 1000 sites of intelligent life within 1000 light years. If we conservatively extend this argument, estimating that only 1 million civilizations with a technology at least equal to ours are distributed throughout the Galaxy, the average separation between them would require 600 years either to send or to receive a message, hardly a hurried conversation.

In order to estimate the number of extraterrestrial civilizations in our Galaxy now, we need a definition of the term. We shall define an extraterrestrial civilization as a group of lifeforms technologically capable of and inclined by curiosity to communication with other Galactic civilizations. We can start by making an inquiry into the possible number of communicative civilizations now existing in our Galaxy. An equation developed by Frank Drake (b. 1930), known as the Drake equation, is a general formula expressing the number of such Galactic communities in terms of several factors:

Number of communicative societies = (astronomical factors) x (biological factors) x (sociological factors)

$$N = R_s \, f_p \, n_p \, f_l \, f_i \, f_c \, L.$$

The first of the astronomical factors, R_s, is the number of stars in our Galaxy divided by the life span of the Galaxy. This factor amounts to about 400 billion stars divided by 10 billion years, or about 40 stars per year. It is a rough measure of the rate at which stars form in the Galaxy.

The second astronomical factor, f_p, is the fraction of stars that live long enough for life to develop and to have a planetary system in which it can develop. From our earlier arguments, about 100 billion main sequence stars live long enough for life to develop, but we assume only half of to have planetary systems, or about 50 billion stars. Therefore, f_p is about 0.125, or 50 billion divided by 400 billion.

The third astronomical factor, n_p, is the number of planets in each planetary system suitable for life; it is the product of the average number of planets per planetary system and the fraction that are suitable for life. For the Solar System, that product equals 1. Therefore, let us somewhat arbitrarily choose one planet for each system.

The first of the biological factors, f_l, is the fraction of planetary systems in which life actually appears. The factor $f_l=0.5$ is arrived at on the assumption that under proper conditions, sooner or later life will take hold, flourish, and evolve into a myriad of thriving forms in every other system.

The second of the biological factors, f_i, is the fraction of evolving systems that evolve at least one intelligent species. We guess that the probability that nature, with say 4 billion years of effort, will create at least one intelligent species on a planet is 50 percent. Therefore, we set the factor f_i equal to 0.5.

Now for the sociological factors. The factor f_c is the fraction of Galactic societies technologically able and willing to take part in interstellar communications; this factor we also guess to be 0.5, that is, a 50 percent chance that the species will develop technological capability and will want to try communicating with other Galactic civilizations.

The factor L is the length of time the civilization continues in its communicative phase. Our own interest in interstellar communication dates back only a few decades in a period of more than 6000 years of civilization.

For the number of intelligent communicative societies, then recognizing a great deal of uncertainty exists in each term, we obtain:

$$N = (40)(0.125)(1)(0.5)(0.5)(0.5)(L)$$

or

$$N = 0.625L.$$

Thus, N is approximately equal to L itself. In other words, the number of communicative civilizations in our Galaxy approximates the average number of years spent in the communicative phase, and the factor L is probably the most uncertain of all to evaluate. Although one may wish to question the assumptions leading to the particular values quoted above, most astronomers believe that they are reasonable. We need to remember also that this result applies only to our Galaxy and does not include the billions upon billions of other galaxies in the Universe.

When we think about the possibilities of our own destruction by nuclear holocaust, by biological disasters from new mutant strains, by changes in the planet's ecology and climatology due to human stupidity and blunders, by terrestrial and extraterrestrial catastrophes, and by other calamities that could befall a civilized society, it is tempting to predict that the moment of civilized glory may indeed be brief in the span of an intelligent species.

Assuming that we have a fair grasp of the values for the product $R_s f_p n_p$, we can calculate the average separation between communicative societies for various values of the product $f_l f_i f_c$ and L. So what about values that cover a range of reasonable values for the product $f_l f_i f_c$ and L? Note that if a species survives in a communicative phase for only 1000 years, then the length of time for messages to travel the distance between civilizations exceeds the lifetime of the communicative phase. We need to look at only the combinations that present any reasonable chance for an exchange of messages.

Radio techniques had improved so spectacularly after World War II that a few astronomers and physicists privately considered the feasibility of detecting extraterrestrial signals from intelligent life. The subject finally surfaced in the British scientific journal Nature in September of 1959, when the physicists Giuseppe Cocconi

(1914-2008) and Philip Morrison (1915-2005) presented logical reasons why efforts should be made to search for interstellar signals generated by intelligent life.

For now it seems more practical for us to listen for signals than to transmit them. Perhaps messages that older, more advanced extraterrestrial civilizations, have been transmitting for centuries have by now reached the Solar System. The most advanced celestial communities could avail themselves of energy sources far more sophisticated and powerful than any we can realize today, perhaps even using the energy output of their parent stars by modulating their light as signals. We may be no more aware of such electromagnetic messages than New Guinea aborigines, who use drums for communication, are aware of the international radio traffic constantly passing overhead.

The first modern attempt in the United States to detect artificial signals from space was conducted by Frank Drake at the National Radio Astronomy Observatory at Green Bank, West Virginia. This undertaking was called Project Ozma, after the legendary princess of the imaginary land of Oz. A 26m radio telescope was aimed at Tau Ceti and Epsilon Eridani for 150 hours of observation from May through July of 1960. Although, the effort was not rewarded by finding signals from intelligent beings, the lack of success seems to have nowise diminished interest. For since then, other attempts to detect signals from intelligent beings have been carried out. None has succeeded, but the several hundred stars examined are a very tiny sample of the possible sources.

The problem in locating signals is to pick not only the right star but also the right frequency and the right time to observe. Even if a communication were received from another world, it would take great amounts of time to exchange messages, so perhaps the first step in acknowledging contact with an extraterrestrial society would be to transmit a duplicate of the received message back to its source to inform the sending society that its inquiry had been received and recognized as originating from an intelligent source.

The groundwork has been laid for developing the search strategy that could ultimately bring us into communication with extraterrestrial civilizations. This program is known as Search for Extraterrestrial Intelligence (SETI). Working on the premise that a large and expensive radio receiving system is not needed to begin with, SETI would equip existing radio telescopes with low-cost state-of-the-art receiving, data-handling, and data-processing equipment. With this apparatus it should be possible to explore the vicinity of the Sun out to several hundred light years for radio leakage from an extraterrestrial civilization or for signals intentionally beamed toward us.

SETI is far more than a single effort. Like the voyages of exploration that discovered the New World or the present missions of planetary exploration, the search would involve many distinct projects with definite goals in mind. These would initially be carried out along with other astronomical investigations, but a time would come when dedicated facilities would be needed. It is estimated that a facility consisting of the collecting area equivalent to a few 100m radio telescopes and associated data-processing equipment could carry out the initial phases of the search. In the event of a positive result or strong prospects for a positive result, a more ambitious program for SETI would be needed.

There are those who believe that we should not even try to contact extraterrestrial intelligent civilizations. In fact, in 2010, one of the premier physicists of the world, Stephen Hawking (b. 1942) elaborated upon this idea. He expects extraterrestrials to be malevolent, like Columbus, and so we should not send any signal alerting extraterrestrials to our existence. However, a group of scholars (including this author), led by Douglas Vakoch (b.

1961) of the SETI Institute, has vehemently denied the Hawking hypothesis that extraterrestrials would be malevolent and have recently published a volume on what is called extraterrestrial altruism. In the end, the point may be quite moot as radio signals from this planet have been travelling into interstellar space for some time now, at least one hundred years.

11.3. Traveling the Universe

The next stage after communication, perhaps carried on simultaneously, would be manned interplanetary flight, that is, travel within the Solar System. The stars are so far from us and the technological and biological difficulties in traveling to them are so great that a trip even to the closest star seems hopeless now. Just consider the time it would take, beginning with a modest round trip to Alpha Centauri, our nearest neighbor: The distance is 4.3 light years; our ship's speed is constant at 50km/s, slightly more than we need to escape from the Solar System, at Earth's distance from the Sun. Our round trip will take about 52,000 years.

If this kind of time scale for space travel is unacceptable, then how then do we get around time? The alternative is to use a ship that can move at something approaching the velocity of light, say 95% of it. This will certainly shorten the trip and also let us take advantage of Einstein's relativistic time dilation. The implications of time dilation for space travel can be illustrated by the following example.

Astronaut A leaves the Earth on a round-trip flight to some star S, 12 light years away, at a speed of 60% the velocity of light. At the same time, astronaut B is travels in the opposite direction on a round-trip voyage to another star T, also 12 light years away, at a speed of 80% the velocity of light. A third person, C remains on Earth to monitor their flights. Before takeoff, the three individuals synchronize their clocks. To avoid complications in recording the traveler's clock times, we assume that the periods of acceleration at the beginning of the outward and return voyages and the periods of deceleration in approaching each star or Earth are extremely brief compared with the time spent in moving at constant velocity. These brief accelerations may therefore be neglected.

It takes light 24 years to make the round trip between Earth and either star S or T. Astronaut A, traveling at 0.6c, will make the trip in 40 years, judged by C's clock, whereas B, traveling at 0.8c, completes the trip in 30 years, again according to C's clock. But A's clock runs slow, so A's round-trip time will be 32 years according to A's clock. B's clock also runs slow, by a different time dilation factor, and hence B's time will be 18 years according to B's clock.

Now suppose all three individuals are 20 years old at the start of their voyage. When B returns, she finds that C is 50 years old, while she is only 38 years old. When A returns, C is 60 years old, and A is only 52 years old. When A and B meet again on Earth 40 years later, A will be only 4 years older than B, because B (38+10 = 48) came back sooner than A did by 10 Earth years. Because biological aging includes a measure of time in molecular cell growth, we presume the ages here are biologically correct.

The recipe for living forever is not simply to move in one direction at a speed close to the velocity of light. An observer who leaves an inertial frame of reference must return to it in order to collect the benefits of the fact that his or her time lags behind that of the observer who remains in the inertial frame. And while in the new frame of reference the biological clock of the body may run slower, so do all bodily activities, so that the pace of living is perceived to be the same.

From the results of the preceding example we can refigure the trip to Alpha Centauri discussed earlier. If we could accelerate the spaceship continuously at a constant rate of 1g, that is, equivalent to the acceleration of gravity that we experience on Earth's surface, we should feel no great discomfort.

The best technique then would be to reverse the acceleration halfway out and come to rest in the vicinity of Alpha Centauri and then on the return trip accelerate at 1g to the halfway point and decelerate at 1g the rest of the way to Earth. Taking acceleration and deceleration into account, the round trip by Earth clocks would be 12 years, but the contracted time aboard the spacecraft, with a maximum speed of 95 percent of the velocity of light, would be 7.2 years.

Thanks to Einstein's Relativity Theory, a journey to the Andromeda galaxy could theoretically be made during an astronaut's life span. Unfortunately, the power a spacecraft would need for an even less ambitious interstellar flight is overwhelmingly large. In fact, it should be noted that relativistic effects with respect to the interstellar medium would subject any travelers, moving at velocities approaching the speed of light, to radiation of energies in the gamma-ray portion of the electromagnetic spectrum. We would need a lead shield at least 10cm (that's about 4 inches) thick to protect us from such radiation, and this would add a prohibitive amount of mass to our spacecraft.

On June 24, 1947, near Mount Rainier, Washington, a salesman named Kenneth Arnold, while flying his own plane, claimed to have seen nine crescent-shaped disks flying near the mountain. His report opened the modern era of "flying saucer," or unidentified flying object (UFO), reports. Arnold's report was neither the first nor the last in that year, but it is the one that caught the attention of the news services. Since that news event in 1947, there have been thousands more in the world press and something in excess of 100,000 UFO reports that never became major news events.

In reality, the UFO "phenomenon" consists of reports of sightings and not the sightings themselves, since for most sightings there is no way of knowing what was observed, if anything. Many sightings, even those with "photographic evidence," have later been admitted to be hoaxes. Although the majority of sightings are by honest and sincere persons, where large volumes of reports have been studied, about 95 percent have been easily attributed to misidentification of "natural" phenomena. This includes airplanes, weather balloons, artificial satellites, planets, bright stars, meteors, ball lightning, flocks of birds, clouds, reflected lights, and luminous insects. The planet Venus itself has prompted more reports than any other cause.

As for the reports that cannot be readily attributed to natural phenomena, the degree of strangeness in a report seems to be inversely correlated with its credibility. That is, where the strangeness is high, such as supposed extraterrestrials waving at the observer, the credibility is low. For example, only one witness saw the event or lighting conditions were extremely bad. There are virtually no high-strangeness, high-credibility reports.

If Earth is not now being visited by extraterrestrials, is it possible that extraterrestrial explorers happened this way in the past and left some sign of their visit? The sign could be some inanimate object or marking, interaction with primitive peoples if they existed, or even we ourselves if the visitors decided to seed the Earth with intelligent life. Yes, this is a possibility. Then, are any of the purported signs reasonable evidence for a visit in the past by extraterrestrial intelligence? They are probably not. For example, the "pictures of space gods," from murals, rock paintings, and pottery figures purported to show extraterrestrial visitors are indeed strange by today's standards. But the anthropologic picture of the culture in question can and does provide a cultural context for the putative space god, so that it is neither necessary nor reasonable to resort to an extraterrestrial visitor as an explanation.

Although it may seem that we are unduly harsh on the possibility of a visit by extraterrestrial beings, this is not truly the case. We find nothing that says that a visit is impossible, in the past, now, or in the future; but what we find woefully unconvincing is the supposed evidence that it is occurring now or has occurred in the past. If it has not already happened, is contact between us and other intelligent beings possible in the future? If they exist, we think it possible at some point. However, there are some very large ifs between now and that first meeting.

For the sake of discussion, assume that it will happen. Then will it be a chance meeting or prearranged? We find it hard to believe that intelligent beings will simply drop by. Instead, it is more probable that the encounter will arrive through a long sequence of message exchanges and that we would have been learning from those beings for a long time before any face-to-face meeting.

11.4. REFERENCES

Breuer, R.A. (1982). **Contact with the Stars**. W.H. Freeman.

Gale, W., ed. (1979). **Life in the Universe**. Westview Press.

Goldsmith, D., ed. (1980). **The Quest for Extraterrestrial Life**. University Science Books.

Goldsmith, D. and T. Owen (1980). **The Search for Life in the Universe**. Benjamin/Cummings.

Vakoch, D.A., ed. (2014). **Extraterrestrial Altruism: Evolution and Ethics in the Cosmos**. Springer-Verlag.

Epilogue - Exploring the Universe
By Harold A. Geller

The place to end our discussions of the Universe is, once again, with us. Our species, Homo sapiens, actually began to explore the Universe some 60,000 years ago, when our ancestors first began the exploration of this planet, coming out of Africa. Homo sapiens migrated to Europe and Asia, then to the Americas and the South Pacific. Our species, in the form of Robert Peary (1856-1920) and Matthew Henson (1866-1955), first reached the North Pole in 1909. The South Pole was reached in 1911 by Roald Amundsen (1872-1928). Today, all over the surface of the Earth, Homo sapiens number over 7 billion.

From the surface of the Earth our species next began to explore what lies beyond. We may consider our exploration of space to actually have begun in 1783 when Joseph Michel Montgolfier (1740-1810) and his brother Jacque Etienne Montgolfier (1745-1799) took the first recorded hot-air balloon journey over Paris, France. With the first sustained controlled flight of an airplane in 1903 by Orville Wright (1871-1948) and his brother Wilbur Wright (1867-1912), the exploration of near space continued and flourished further.

The exploration of space with rockets blossomed with Robert Goddard (1882-1945) and Wernher von Braun (1912-1977). The first satellite to orbit the Earth was launched by Russia in 1957. It started a "space race" which led to the first members of our species to walk on the surface of the Moon, in the form of Neil Armstrong (1930-2012) and Edwin "Buzz" Aldrin (b. 1930).

Summarizing our exploration of the Universe, we find that there have been some 72 missions to the Moon, our nearest sizeable cosmic neighbor. There have been two missions to Mercury and some 40 missions to Venus. Mars has been the object of study for some 38 missions. Jupiter has seen some 8 missions, with Juno being the most recent on its way to an encounter in 2016. Saturn has been the target for 5 missions, including a probe that landed on the surface of its moon Titan, the only satellite in our Solar System with a substantial atmosphere. Uranus and Neptune have only been visited by human built probes once, in the guise of Voyager 2. Pluto is yet to be visited by the New Horizons mission, set to visit Pluto in the summer of 2015.

As of now, there are only five spacecraft built by our species that will ever leave the Solar System. These are the Pioneer 10, Pioneer 11, Voyager 1, Voyager 2, and New Horizons spacecraft. Voyager 1 is currently the most distant, and now the first of any ever built by our species to leave the confines of our Sun and enter true interstellar space. Will humans ever again venture off this planet and continue our exploration of the Universe with our own bodies? Who reading this book will be able to say that a spacecraft containing some samples of our species will be heading to another star system and continue human exploration somewhere beyond the sea of interstellar space?

About the Editor

Dr. Harold A. Geller is Observatory Director and Associate Professor at George Mason University (GMU). He is co-Investigator for the Virginia Initiative for Science Teaching and Achievement. He has served as: Solar System Ambassador, NASA Jet Propulsion Laboratory; Associate Chair, GMU Physics and Astronomy Department; President, Potomac Geophysical Society; award winning tour guide at NASA Goddard Space Flight Center; producer of educational multimedia CD-ROMs; faculty at Northern Virginia Community College; doctoral fellow, State Council of Higher Education for Virginia; and, planetarium operator, National Air and Space Museum. He won six Telly Awards (with Astrocast.TV) and Faculty Member of the Year Award. He is author of: four books and over 75 papers in biochemistry, education, astrobiology, and astrophysics. He has been interviewed or quoted in the USA Today, Washington Post, Huffington Post, Arts and Entertainment Magazine, Astrocast TV, WTOP News Radio, News Channel 8, and The Skeptic.

About the Illustrator

Emma Rojas was an administrative assistant and graphic artist for the department of Geography and GeoInformation Sciences (GGS) at George Mason University (GMU). She was an undergraduate major in Foreign Languages with a concentration in Spanish. She was also minoring in Geography. Emma received her Bachelor of Arts degree from George Mason University in May of 2015. She designed various informational graphics for educational purposes; illustrated flyers, posters, brochures, logos, bookmarks, and floor plans; has experience in creating, updating and maintaining content on webpages. Her illustrations have appeared in university brochures and posters; and, presentations for professional organizations.

www.ingramcontent.com/pod-product-compliance
Lightning Source LLC
Chambersburg PA
CBHW031826170526
45157CB00001B/202